AF773382

IN SITU URANIUM STABILIZATION THROUGH POLYPHOSPHATE REMEDIATION

IN SITU URANIUM STABILIZATION THROUGH POLYPHOSPHATE REMEDIATION

D. M. WELLMAN, E. M. PIERCE,
V. R. VERMEUL, S. V. MATTIGOD,
E. L. RICHARDS, M. D. WILLIAMS,
J. S. FRUCHTER AND J. P. ICENHOWER

Nova Science Publishers, Inc.
New York

For permission to use material from this book please contact us:
Telephone 631-231-7269; Fax 631-231-8175
Web Site: http://www.novapublishers.com

NOTICE TO THE READER

Library of Congress Cataloging-in-Publication Data

In situ uranium stabilization through polyphosphate remediation / D.M. Wellman ... [et al.].
p. cm.
ISBN 978-1-60456-940-7 (softcover)
1. Radioactive waste disposal. 2. Uranium ores. 3. Phosphate rock. I. Wellman, D. M. (Dawn M.)
TD898.I48 2008
621.48'38--dc22

2008025978

Published by Nova Science Publishers, Inc. ✣ New York

ACKNOWLEDGMENTS

This work was conducted at Pacific Northwest National Laboratory, operated by Battele Memorial Institute for the U.S. Department of Energy under Contract DE-AC05-76RL01830. Funding for this project was provided by the U.S. Department of Energy, Office of Environmental Management, EM-20 Environmental Cleanup and Acceleration, Mark Gilbertson; and the Enfironmental Remediation Sciences Program, Office of Biological and Enrironmental Research. We appreciate the support of Richland Operations Office, Mike Thompson. We thank S.D. Burton, B.C. Butler (Stanford University), B.K. McNamara B.W. Arey, R.N. Ermi, J.N. Glovack (Colorado School of Mines), K.E. Parker and E.A. Rodriguez for their support and operation of laboratory testing. We Thank B.A. Williams, D.R. Newcomer, B.G. Fritz, R.D. Mackley, D.P. Mendoza, K.M. Gunderson, and K.N. Watts for their support of field activities. The assistance of E.T. Clayton for conducting ICP-MS, S.R. Baum for ICP-OES, and Michelle M. Valenta for IC analyses is greatly appreciated.

CONTENTS

PREFACE

The Hanford Site, in southeastern Washington State, is a former nuclear defense production facility. A groundwater plume containing uranium, originating from a combination of purposeful discharges of wastewater to cribs, trenches, and ponds, along with some accidental leaks and spills with nuclear fuel fabrication activities, has persisted beneath the Hanford Site 300 Area for many years. The uranium plume is just upstream of the city of Richland municipal water supply intake on the Columbia River. Despite the cessation of uranium releases and the removal of shallow vadose zone source materials, the remedial action objective to lower the concentration of groundwater uranium to the U.S. Environmental Protection Agency (EPA) maximum contaminant level (MCL) concentration of 30 µg/L has not been achieved within the anticipated 10-year time period.

Despite several decades of studies, effective uranium cleanup strategies remain elusive for contamination in deep subsurface settings that prevail at a number of U.S. Department of Energy (DOE) sites in the western United States. Numerous strategies have been proposed including iron barriers, soluble reductive agents, microbial stabilization via reduction and precipitation, and emplacement of solid phosphate barriers. While these all have merit, each encounters limitations for deep subsurface remediation in an oxidative environment (Peterson et al. 2005; Wellman et al. 2005b; Wellman et al. 2006c; McGrail and Mattigod 1999; DOE 2005a; DOE. 2005b). Polyphosphate technology has been demonstrated to delay the precipitation of phosphate phases for controlled in situ formation of stabile phosphate phases which control the long-term fate of uranium (Wellman et al. 2005b; Wellman et al. 2006c). Precipitation of phosphate minerals occurs when phosphate compounds degrade in water to yield the orthophosphate molecule (PO43). Accordingly, a detailed understanding of polyphosphate degradation and reaction kinetics, in the context of site-specific information, allows the technology to be tailored as a time-released source of phosphate for direct subsurface treatment without a drastic change in hydraulic conductivity of the target aquifer.

A site-specific treatability test was conducted to optimize polyphosphate remediation technology for implementation through a pilot-scale field technology demonstration to accelerate monitored natural attenuation of the uranium plume within the Hanford 300 Area aquifer. A focused application of polyphosphate was conducted in a source or "hot spot" area to reduce the inventory of available uranium that contributes to the groundwater plume through direct precipitation of uranyl-phosphate solids and secondary containment via precipitation of apatite acting as a long-term sorbent for uranium. The general treatability testing approach consisted of initial site characterization and setup, a polyphosphate injection test, and post-treatment performance assessment. Fundamental science studies were conducted with site- specific sediment and groundwater to develop an effective remediation scheme for deployment of polyphosphate technology. In addition to remediating a portion of the plume, the data from this test provide valuable information for designing a full-scale remediation of uranium in the aquifer at the 300 Area of the Hanford Site. Results provide a detailed understanding of the fundamental underpinnings necessary to evaluate the efficacy and potential use of polyphosphate technology at other sites with varying geochemical and hydrodynamic conditions.

Chapter 1

BACKGROUND

HANFORD SITE AND THE 300 AREA URANIUM PLUME

The Hanford Site, located in the eastern part of Washington State, contains nuclear facilities operated by the U.S. Department of Energy (DOE) (figure 1). These facilities were operated from 1944 to 1989 to produce fissionable nuclear material. In 1989, all nuclear reactors were shut down and the activities were shifted towards environmental cleanup and site restoration. During the period spanning the startup of Hanford reactors in 1944 through the late 1980s, facilities in the 300 Area of the Hanford Site were primarily involved with the fabrication of nuclear fuel and the range of activities associated with this task resulted in a wide variety of waste streams that contained chemical and radiological constituents (Gerber 1992; DeFord et al. 1994). Since the early 1990s, extensive remediation of liquid waste disposal sites and solid waste burial grounds has taken place. Recently, most liquid waste disposal sites, which are located in the north half of the 300 Area (figure 1), have been excavated, backfilled, and the ground surface restored. Some unknown amount of contamination remains in the vadose zone beneath the lower extent of the excavation activities. Additional contamination may also remain beneath buildings and facilities in the southern portion of the 300 Area, where decontamination and decommissioning activities have not yet taken place.

The most prominent waste constituent remaining in this environment is uranium. Groundwater beneath the 300 Area contains uranium from past-practice disposal activities at concentrations that exceed the U.S. Environmental Protection Agency (EPA) standards for drinking water supplies (figure 2). Currently, elevated uranium concentrations are entering the Columbia River along the shoreline and enter the riparian and river biota through seeps. Uranium in its soluble form is of concern because of its chemical toxicity and risk of radiological exposure, even though the concentrations in groundwater for chemical toxicity are lower than those associated with exceeding radiological dose standards.

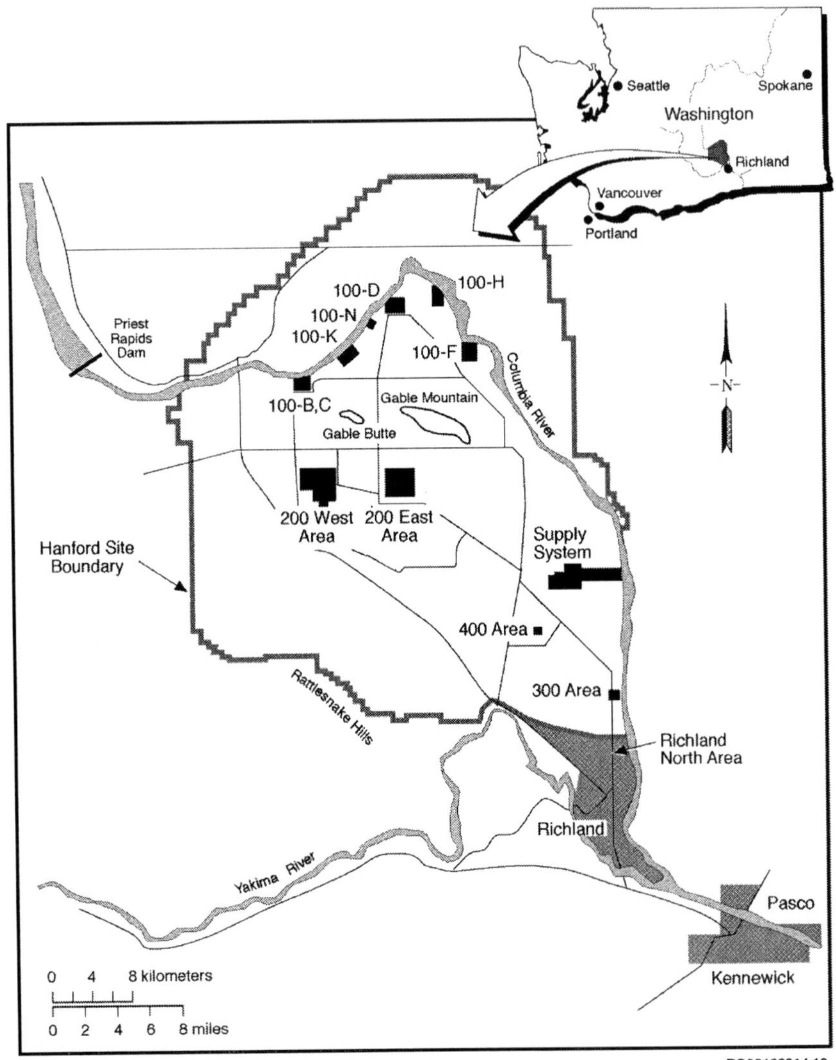

Figure 1. Map of the Hanford Site.

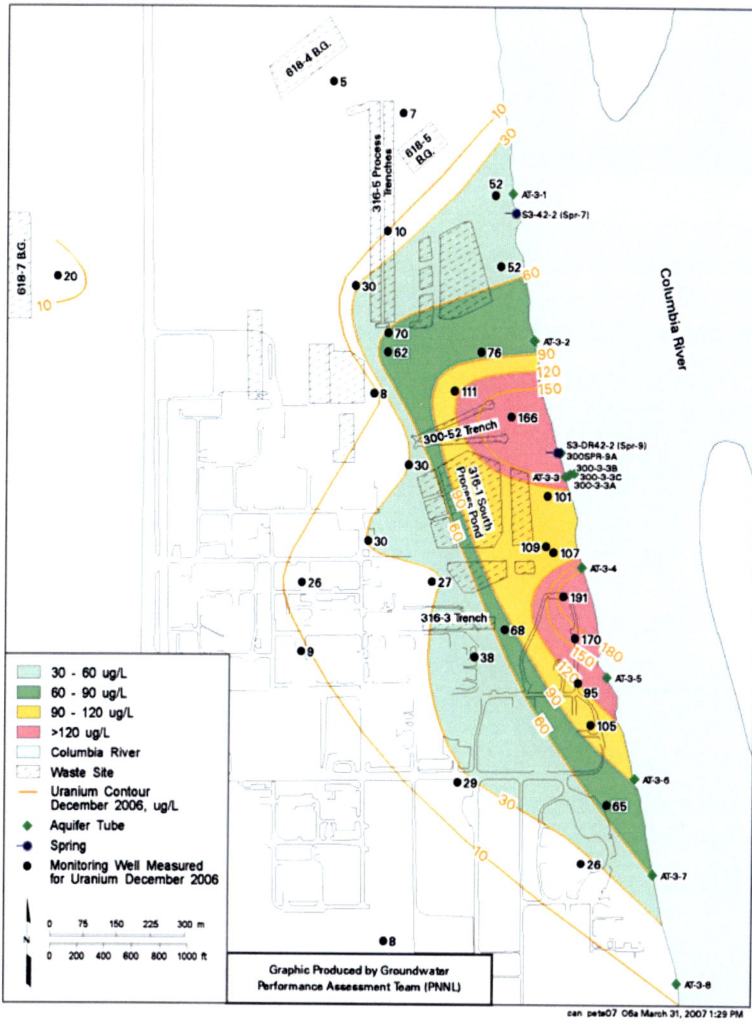

Figure 2. Schematic Depicting the Concentration Contours of the Uranium Plume Within the Hanford Site 300 Area Aquifer as of December 2006.

In 1996, a record of decision (ROD) for the Hanford 300 Area stipulated an interim action program of a natural attenuation process accompanied by increased groundwater monitoring (DOE 1995; EPA 1996). During the remedial investigation in the early 1990s and the development of the initial ROD, the proposed standard for uranium was fixed at 20 μg/L as compared to the current EPA maximum contaminant level (MCL) of 30 μg/L.

Despite the cessation of uranium releases and the removal of shallow vadose zone source materials, dissolved uranium concentrations below the cleanup criteria established by the ROD have not been achieved within the anticipated 10-year time period. Therefore, a feasibility study was begun in 2005 to identify and evaluate remedial alternatives that will accelerate monitored natural attenuation of the uranium plume.

Among the several remediation technology alternatives, polyphosphate application was judged to be the most promising technology to sequester uranium at this site. The technology works by forming phosphate minerals (autunite and apatite) in the aquifer, which directly sequesters the existing aqueous uranium in autunite minerals and precipitates apatite minerals for sorption and long-term treatment of uranium migrating into the treatment zone, thus reducing current and future aqueous uranium concentrations (Wellman et al. 2005b, 2006c). In this study, we will describe the intermediate-scale bench-scale tests that were conducted to optimize this technology and the preliminary results of a pilot-scale field test that was conducted in 2007 (Wellman et al. 2006a, 2007; Vermuel et al. 2006;).

Uranium Geochemistry and Aqueous Speciation

The most abundant actinide element in the natural environment is uranium, with an average composition of ~1.2 to 1.3 μg/g in sedimentary rocks, ~2.2 to 15 μg/g in granites, and ~20 to 120 μg/g in phosphate rocks. Uranium has four known oxidation states: U(III), U(IV), U(V), and U(VI); however, only two dominate in natural systems. These valance states have a significant affect on the mobility of uranium in subsurface environments. The uranous ion [U(IV)] dominates under reducing conditions, whereas the uranyl ion (UO_2^{2+}) dominates under oxidizing conditions. Under most redox conditions, dissolved U(III) oxidizes to U(IV), while dissolved U(V), aqueous species (UO_2^{+}), disproportionates to U(IV) and U(VI), aqueous species UO_2^{2+}. The mobility of uranium in the aquifers beneath the 300 Area is controlled by a number of different factors, including 1) complexation with anions and cations in the aqueous phase to form aqueous species or complexes; 2) sorption by naturally occurring minerals, primarily oxide/oxyhydroxides; and 3) dissolution from uranium-bearing minerals or phases. Although the latter two processes are important to the transport of uranium, this discussion focuses on the aqueous uranium speciation.

In the oxidized environment of the 300 Area aquifer, dissolved UO_2^{2+} reacts with aqueous anions and cations (i.e., carbonates, chlorides, fluorides, hydroxides, phosphates, and sulfates) to form aqueous complexes. The presence or formation of a particular aqueous species is determined by: 1) the total uranium concentration, 2) the concentration of a particular ligand, 3) the equilibrium or stability constant for a particular aqueous complex, and 4) the solution pH and temperature. Because of the analytical limitations associated with identifying solution species in situ, calculations using geochemical speciation models such as MINTEQA2 (Allison et al. 1991, 1998), EQ3NR (Wolery 1992; Wolery and Daveler 1992; Wolery and Jarek 2003), or equivalent are conducted to obtain a theoretical depiction of uranium aqueous speciation based on thermodynamic considerations. The theoretical foundations for the calculations that are performed with geochemical codes are described in a variety of environmental chemistry and geochemistry text books. The thermodynamic data needed to perform these calculations are derived from experimental results obtained from various literature sources. The quality of these results vary, therefore the values assigned to these species may have significant uncertainty. Detailed literature reviews have been provided by Grenthe et al. (1992) and Guillaumont et al. (2003), which are compilations of the most widely accepted literature regarding uranium thermodynamics and speciation. The calculations discussed in the following figures were performed using these values.

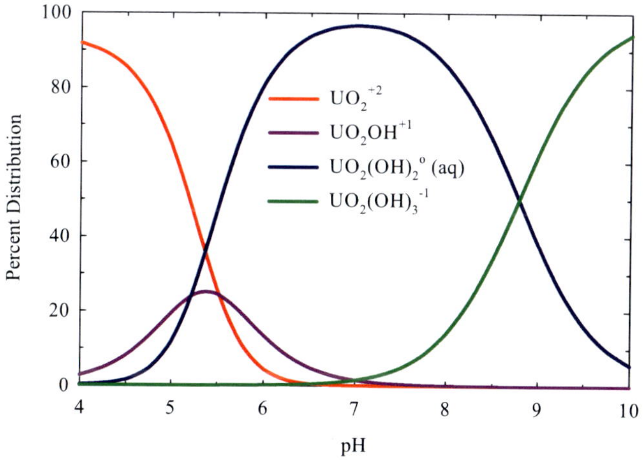

Figure 3. Percent Distribution of U(VI)-H_2O System Calculated with MINTEQA2 at 25°C, I = 0.1 M, and total U(VI) = 1×10^{-6} M in the Absence of Complexing Ligands other than Hydroxide.

In a system free from complexing ligands other than dissolved hydroxides, the UO_2^{2+} hydrolyzes to form a number of uranyl-hydroxide complexes (figure 3). In this simple system four U(VI) complexes, UO_2^{2+}, $UO_2(OH)^+$, $UO_2(OH)_2^\circ$, and $UO_2(OH)_3^-$, are present and the dominant species changes as pH increases from 2.0 to 10.0. The modeling calculations predicted the UO_2^{2+} cation dominates under acidic conditions and gradually decreases as the pH increases from 2.0 to 5.5. The $UO_2(OH)_2^\circ$ complex begins to increase at pH > 5 and becomes the dominant uranium species in solution, from pH > 5.5 to 9.0. Beyond pH > 9.0 the dominant species changes from $UO_2(OH)_2^\circ$ to $UO_2(OH)_3$.

In most natural waters, which are in equilibrium with atmospheric CO_2, the aqueous speciation of uranium changes dramatically in comparison to the U(VI)-H_2O system shown in figure 3. The uranyl ion forms strong carbonate complexes in most natural waters and is illustrated by the modeling calculation performed for the U(VI)-H_2O-CO_2 system (figure 4, top). In this system, the uranyl-hydroxide complexes are redistributed to form mainly uranyl-carbonate complexes; $UO_2(CO_3)_2^{2-}$, $UO_2(CO_3)_3^{4-}$, at pH values greater than 5. These results illustrate that at pH > 7.0 the $UO_2(CO_3)_3^{4-}$ complex becomes dominant. As previously stated, the pH and concentration of complexing cations and anions can play significant roles in which species dominate the system. This is show in figure 4 (bottom), where the percent U(VI) distribution in Hanford groundwater, obtained from well 699-S3-25, has been plotted. The chemical composition of the Hanford Site groundwater (table 1) has been fairly constant over 10 mattyears and is considered to be in equilibrium with the soil environment. The results shown in figure 4 demonstrate the complexity of the U(VI) speciation in Hanford Site groundwaters and it also demonstrates the impact calcium, as the $Ca_2UO_2CO_3$ complex, can have on the speciation of U(VI). This complex is expected to be the dominant U(VI) form under the conditions expected in the Hanford subsurface. Similar to Ca^{2+}, the presence of Mg^{2+} has been recently suggested to also form a $Mg_2UO_2CO_3$ complex. This complex was not included in the calculations displayed in figure 4 due to a lack of thermodynamic data, but the presence of this complex could have a profound effect on the U(VI) distribution, given the concentration of Mg^{2+} in Hanford groundwater (table 1).

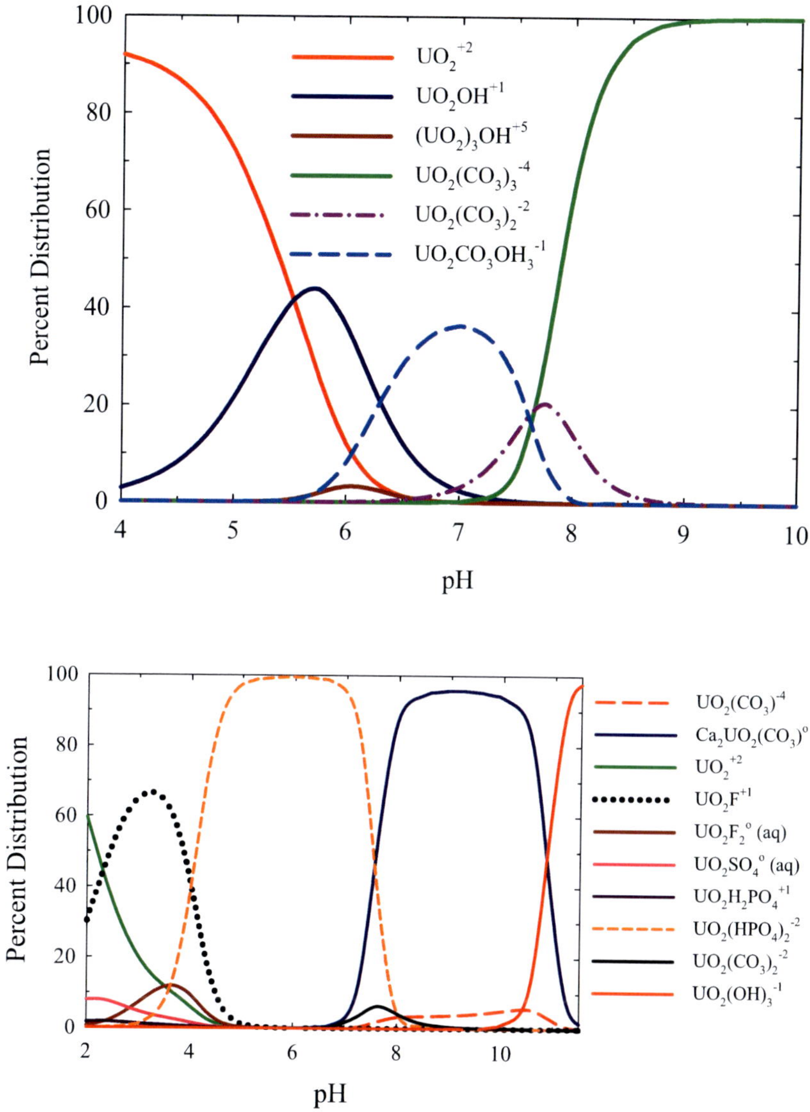

Figure 4. Percent Distribution of U(VI) Species Calculated with MINTEQA2 at 25°C, Ionic Strength = 0.1 M, and $pCO_2 = 10^{-3.5}$ Bar for a Total U(VI) = 1×10^{-6} M (top); Percent Distribution of U(VI) Aqueous Species Calculated with MINTEQA2 in Hanford Groundwater Well-699-S3-25 (bottom).

Table 1. Composition of Typical Hanford Groundwater (HGW) from Monitoring Well-699-S3-25. Data obtained from Serne et al. (1995)

Element	Concentration (mg/L)
Alkalinity (as $CaCO_3$)	140.0
Ba	0.061
Br	0.10
Ca	59.0
CO_3^{2-}	0.80
Cl	21.0
Co	0.0041
F	0.60
Fe	0.22
Mg	15.0
Mn	0.09
NO_3^-	1.40
PO_4^{3-}	<0.34
K	7.70
Ag	0.0031
Na	26.0
SO_4^{3-}	110.0
H_4SiO_4	81.84
U	0.01
Specific Conductance (μS/cm)	557
Temperature (°C)	19.10
pH	7.72
Eh (mV)[(a)]	146.0
Total Hardness as $CaCO_3$ (mg/L)	209.48

Data taken from the Hanford Environmental Information Systems (HEIS) database using Monitoring Well-699-S6-E14A provides a general reference for the Hanford groundwater redox conditions.

Polyphosphate Remediation Technology

Numerous approaches have been proposed to sequester uranium, in situ, with solid-phase hydroxylapatite, $Ca_5(PO_4)_3OH$, (Conca 1996; Arey et al. 1999; Wright et al. 1995; Seaman et al. 2001; Moore et al. 2001; Gauglitz and Holterdorf 1992), and water-soluble phosphate compounds, such as tribasic sodium phosphate [$Na_3(PO_4){\cdot}nH_2O$] (Lee et al. 1995) or phytic acid (Jensen et al. 1996; Nash et al. 1998b, 1998a, 1999), which could be injected into contaminant plumes from strategically placed wells, acting as a chemical stabilizer for uranium and other radionuclides, as well as heavy metals. The advantages of soluble amendments is that they allow for treatment of plumes situated deep within the subsurface and act to sequester uranium by precipitating insoluble uranium minerals rather than by reversible sorption mechanisms or redox manipulations. However, Wellman et al. (2006c) demonstrated that compounds including tribasic sodium phosphate and phytic acid result in the rapid formation of phosphate phases. Formation of these phases occludes ~30% of the fluid-filled pore space within the sedimentary formation. Rapid reduction in the hydraulic conductivity will have a significant effect on subsequently injected amendment solutions, the targeted groundwater plume, or both, by deflecting flow from the natural path.

Conversely, the use of soluble long-chain polyphosphate materials have been demonstrated to delay the precipitation of phosphate phases (Wellman et al. 2005b, 2006c) (figure 5). Precipitation of phosphate minerals occurs when polyphosphate compounds hydrolyze to yield the orthophosphate molecule (PO_4^{3-}), which then binds with metal ions such as uranium. The longer the polyphosphate chain, the slower the hydrolysis reaction leading to orthophosphate production (Shen and Morgan 1973) (figure 6). Accordingly, use of a long-chain polyphosphate compound does not result in a drastic change in hydraulic conductivity of the target aquifer.

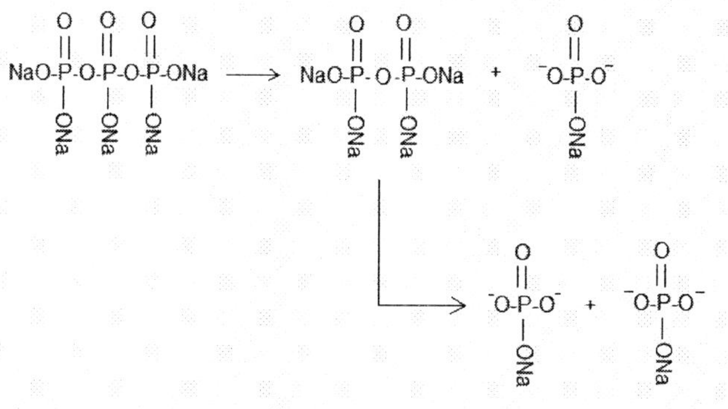

Figure 5. Schematic Depicting the Step-Wise Hydrolysis of Sodium Tripolyphosphate.

Injection of a sodium tripolyphosphate amendment into the uranium-bearing saturated porous media has been shown to immobilize uranium through the formation of an insoluble uranyl phosphate mineral, autunite $X_{1-2}[(UO_2)(PO_4)]_{2-1}\bullet nH_2O$, where X is any monovalent or divalent cation. Because autunite sequesters uranium in the oxidized form, U(VI), rather than forcing reduction to U(IV), the possibility of re-oxidation and subsequent re-mobilization of uranium is prevented.

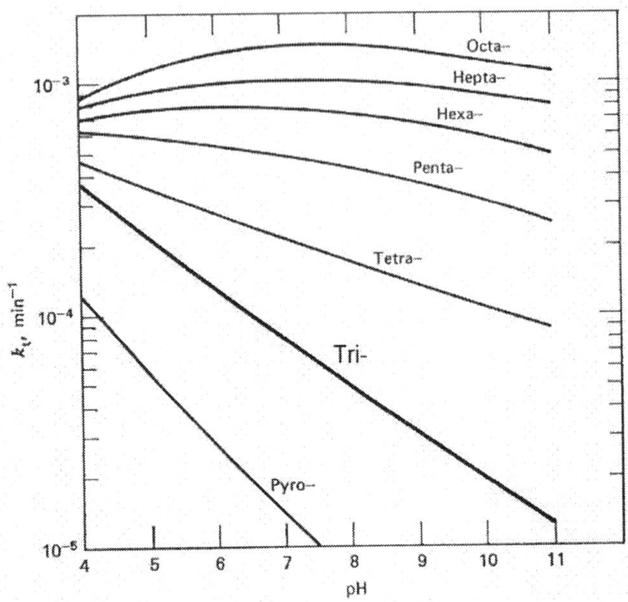

Figure 6. Hydrolysis Rate of Polyphosphate Molecules as a Function of pH (Shen and Morgan 1973).

Release of uranium from autunite may only occur through dissolution of the mineral structure. Extensive testing demonstrates the very low solubility and slow dissolution kinetics of autunite under conditions relevant to the Hanford subsurface (Wellman et al. 2006b). In addition to autunite, excess phosphorous can result in apatite (calcium phosphate) mineral formation, providing a long-term source of treatment capacity.

Research beginning in the mid-1960s underscored the efficacy of using calcium and/or lime to precipitate stable calcium-phosphate solid phases including apatite for direct removal of phosphate (Ferguson et al. 1973, 1970; Schmid and McKinney 1969; Jenkins et al. 1971). By complexing calcium and sorbing to mineral surfaces, polyphosphate compounds effectively enhance the rate of calcium-phosphate precipitation by reducing competing reactions, such as the formation of calcium-carbonate, and "directing" the calcium to participate in reactions resulting in calcium-phosphate precipitation (Ferguson et al. 1973).

Fuller et al. (2002; 2003) demonstrated the efficacy of hydroxylapatite for reducing the aqueous uranium concentration to <0.05 μM under the pH range of 6.3 to 6.9 in the presence of carbonate. Results suggested the binding of uranium, irrespective of dissolved carbonate concentration or aqueous uranium concentration, occurred via surface complexation; long-term retention occurs through the transformation of sorbed apatite to H-autunite (chernikovite). Similar evidence for the long-term retention of uranium via initial sorption and subsequent transformation to uranium mineral phases of low solubility has been observed downgradient from the uranium ore deposit at Koongarra, Australia (Murakami et al. 1997).

The degradation kinetics of long-chain polyphosphate molecules, such as sodium tripolyphosphate, affords a time-released source of inorganic phosphate. This allows controlled application and precipitation within deep subsurface contaminant plumes. However, site-specific optimization of polyphosphate technology is critical to successful implementation of these technologies in the field (Wellman et al. 2006a). A thorough understanding of site-specific geochemical and hydrodynamic conditions enables optimization of polyphosphate technology for controlled remediation without discharging excess phosphorus into the environment, which could detrimentally impact the environment or downgradient drinking water supplies. Research presented here details the optimization and pilot-scale field testing of polyphosphate technology within the 300 Area of the Hanford Site (Wellman et al. 2006a, 2007; Vermuel et al. 2006). Results underscore the necessity to understand the geochemistry controlling uranium within the subsurface environment and evaluate the resulting effect of phosphate-based remediation technology on the system.

Chapter 2

EXPERIMENTAL METHODS FOR POLYPHOSPHATE TECHNOLOGY DEVELOPMENT

POLYPHOSPHATE HYDROLYSIS EXPERIMENTS

The speciation of inorganic phosphate and its chemical affinity for other species in solution can be readily assessed with phosphorus-31 nuclear magnetic resonance (^{31}P NMR). Controlled ^{31}P NMR experiments were conducted to quantify the kinetic degradation rate of the tripolyphosphate molecule under conditions present within the Hanford 300 Area subsurface. The effects of aqueous cations (e.g., Al^{3+}, Ca^{2+}, Fe^{3+}, Mg^{2+}, and Na^{+}) and sedimentary materials (e.g., FeOOH, and native Hanford sediment) on the hydrolysis of polyphosphates were evaluated in potassium carbonate (K_2CO_3) buffered solutions at 23°C. The K_2CO_3 buffer was used to maintain the pH in a range near that of Hanford groundwater (pH = 7.5 to 8.5) because carbonate is a major component of Hanford subsurface and groundwater, and because potassium has been shown to have a low catalytic effect on phosphate hydrolysis (Wieker and Thilo 1960).

Homogeneous hydrolysis experiments were conducted by first preparing buffered metal chloride stock solutions. These solutions were prepared by mixing 700 mL of 0.1 M K_2CO_3 in deuterated water (D_2O) and adjusting the pH by adding 4.86 mL of HCl. Prior to pH-adjusting the influent solutions, the pH probe was calibrated with National Bureau of Standards (NBS) buffers (pH = 7.00, 10.00, and 12.00 at 23°C). The precision of the pH measurement was ±0.02. Once prepared, the stock buffer was then divided into five 100-mL fractions and one 200-mL fraction. To each 100-mL fraction was added 1 mM equivalent of one of the following metal chlorides: $AlCl_3$, $CaCl_2$, $FeCl_3$, and $MgCl_2$. Precipitation occurred in the $AlCl_3$, $CaCl_2$, and $FeCl_3$ stock solutions; probably as $Al(OH)_3$, $CaCO_3$, and $Fe(OH)_3$, respectively. The final dissolved concentrations for aluminum (4.10×10^{-16} M), iron (8.36×10^{-22} M), and calcium (4.0×10^{-8} M) are based upon the solubility limit of the aforementioned phases. All chemicals used in these experiments were reagent grade. Cations used for these experiments are representative of some of the major components of Hanford Site sediment and groundwater. Each homogeneous hydrolysis experiment began by mixing approximately 5 mL of the appropriate buffered metal solution with 0.366 g of solid sodium tripolyphosphate, which corresponded to 0.2 M of tripolyphosphate solution.

Unlike the homogeneous experiments, each heterogenous solid experiment contained 2.5 g of Hanford sediment and 44 mg of FeOOH per 5 mL of 0.2 M sodium tripolyphosphate carbonate buffered solution. Each experiment was sampled weekly for four weeks. Approximately 1.5 mL of sample was removed, filtered, and placed in a 5-mm outer-diameter thin-walled precision Wilmad® glass NMR tube and analyzed immediately.

Non-proton decoupled ^{31}P NMR spectra were recored on a two-channel Varian-VXR, operating at 300-MHz proton frequency (i.e., 7.0T). A 4.5-μsec 90° pulse was used with a 0.5-sec pulse delay, 1.813-sec acquisition time, a frequency of 121.43 MHz, and 300 acquisitions per sample. Spectra were referenced to the resonance peak of 85% phosphoric acid (H_3PO_4, 37.9 ppm), which was used as an external chemical shift standard. All ^{31}P NMR experiments were conducted at room temperature in D_2O.

Analysis of the D_2O/phosphate controls, which should have yielded constant peak areas over the course of the experiment (Willard et al. 1975; Wazer 1958), indicated there were fluctuations in the measured areas. This artifact was also reflected in the analysis of the experimental results. However, the ratios of the peak areas within the standards were constant. This consistency implied the peak areas of each sample were being influenced by an analytical artifact that could be correlated to a constant determined from the D_2O/phosphate controls and used to "scale" the sample peak, thus allowing a single set of standard curves to be used to calculate concentration information for each sample.

Peak areas for all phosphate species were computed and tallied. The standard curves indicated the phosphate per peak area ratio was different for each phosphate species. Therefore, ratios of the slopes of the standard curves were used to normalize all peak areas to a single concentration per peak area ratio. Each peak area was then multiplied by the number of phosphate molecules contained in the corresponding species to yield a single phosphate per peak area ratio. Peak areas were then divided by the sum of all peak areas in the sample to yield the percent total dissolved phosphate represented by each species. For tripolyphosphate, the doublet area was used instead of the triplet area because of its greater size and differentiation from other peaks.

Chapter 3

AUTUNITE AND APATITE FORMATION

BATCH EXPERIMENTS – AMENDMENT FORMULATION AND EFFICACY

Batch experiments were conducted over a range of polyphosphate sources and concentrations to identify the required conditions to obtain maximum precipitation of autunite and/or apatite (table 2). Selection of the amendment sources was based on the solubility, hydrolysis rate, and amount of phosphorus and/or calcium provided by the respective compounds. The thermodynamic geochemical code EQ3NR (Wolery 1992) was used to assess the necessary concentration of phosphorus to precipitate hydroxylapatite and autunite given the minimum and maximum saturation state measured within the aquifer. Thermodynamic databases from numerous literature sources were used to update the computer code (Sergeyeva et al. 1972; Langmuir 1978; Alwan and Williams 1980; O'Hare et al. 1976, 1988; Vochten 1990; Nguyen et al. 1992; Grenthe et al. 1992; Finch 1997; Chen et al. 1999; Kalmykov and Choppin 2000). It is important to note that because of the complex chemistry of uranium, there is significant debate within the literature regarding the stoichiometry and the thermodynamic values assigned to aqueous uranium species and secondary mineral phases. As such, the geochemical calculations are based on current knowledge, but may have significant uncertainty associated with them.

Batch tests were conducted using a solid-to-solution ratio of 1 to 100. All experiments were conducted in Hanford groundwater and in the presence of 300 Area sediments for one week at room temperature. The uranium concentrations used in these experiments (10 to 1000 ppb) were selected to represent the range of concentrations measured within the Hanford 300 Area aquifer. Aqueous concentrations were monitored by inductively coupled plasma-mass and optical emission spectrometries (ICP-MS and ICP-OES).

Table 2. Proposed Phosphate Sources for Polyphosphate Amendment

Phosphate Source	Formula	Ca Source	P Conc (ppm)	Ca Conc (ppm)
Sodium Orthophosphate	$Na_3PO_4 \cdot nH_2O$	$CaCl_2$	500, 1000, 1500, 2000, 2500	0, 1000, 1500
Sodium Pyrophosphate	$Na_4P_2O_7 \cdot nH_2O$	$CaCl_2$	500, 1000, 1500, 2000, 2500	0, 1000, 1500
Sodium Tripolyphosphate	$Na_5P_3O_{10} \cdot nH_2O$	$CaCl_2$	500, 1000, 1500, 2000, 2500	0, 1000, 1500
Sodium Trimetaphosphate	$(NaPO_3)_3 \cdot nH_2O$	$CaCl_2$	500, 1000, 1500, 2000, 2500	0, 1000, 1500
Sodium Hexametaphosphate	$(NaPO_3)_6 \cdot nH_2O$	$CaCl_2$	500, 1000, 1500, 2000, 2500	0, 1000, 1500
Calcium Dihydrogen Phosphate	$Ca(H_2PO_4)_2 \cdot nH_2O$	–	500, 1000, 1500	–
Calcium Hydrogen Phosphate	$CaHPO_4 \cdot nH_2O$	–	500, 1000, 1500	–
Calcium Pyrophosphate	$Ca_2P_2O_7 \cdot nH_2O$	–	500, 1000, 1500	–
Calcium Hypophosphite	$Ca(H_2PO_2)_2 \cdot nH_2O$	–	500, 1000, 1500	–

Column Testing - Amendment Formulation, Efficacy, and Emplacement

Site-specific tests were conducted to optimize the formation of apatite based on environmental parameters, including pH and carbonate concentration. Saturated column experiments were conducted to quantify 1) polyphosphate treatment efficiency—the amount of polyphosphate required to treat a pore volume of uranium contaminated groundwater, and 2) polyphosphate treatment emplacement efficiency—to evaluate the mixing problem (i.e., effective contact or tendency for the reagent to push contaminated groundwater ahead of the treatment volume).

The use of multi-length polyphosphate chain amendments, optimized through ^{31}P NMR hydrolysis and batch precipitation experiments (ortho-, pyro-, and tripolyphosphate), was evaluated to afford rapid precipitation of autunite and/or apatite without negatively impacting the hydraulic conductivity of the formation. Preliminary characterization of geologic media collected from the 300 Area indicated that the uranium concentration within the aqueous and solid matrix of the sediment cores is below the MCL for uranium. Therefore, to effectively evaluate polyphosphate amendments for uranium remediation, it was necessary to use a solution of Hanford groundwater spiked with aqueous uranium as the influent solution, with a maximum concentration in the pore fluid of 1000 ppb.

Polyvinyl chloride columns (length, L = 30.48 cm; radius, r = 2.54 cm; and bulk volume, V_b = 194.04 to 202.20 cm^3) were uniformly packed with saturated sediment from the 300 Area. These flow tests were conducted to evaluate the concentration of total phosphorus and calcium; the ratio of ortho-, pyro-, and tripolyphosphate; the molar ratio of calcium to phosphorus; pH; and the injection order to optimize emplacement of the amendment and the extent of treatment, reduction in aqueous uranium concentration, and the formation of autunite and apatite. Sodium orthophosphate ($Na_3PO_4 \cdot 12H_2O$), sodium pyrophosphate ($Na_4P_2O_7 \cdot 10H_2O$), and sodium tripolyphosphate ($Na_5P_3O_{10}$) provided the source of each respective phosphate for all phosphorus amendment formulations and calcium-chloride ($CaCl_2$) as the source of calcium. The polyphosphate amendment was contained in one solution and the calcium amendment composed a separate solution. Calcium rapidly precipitates with orthophosphate; therefore, all injections were conducted in two phases by injecting either the calcium solution followed by the phosphorus solution or vice versa. Details regarding the amendment formulation, injection order, calcium-to-total phosphorus molar ratio, amendment pH, and concentrations are summarized in table 3.

Following saturation and attainment of chemical equilibrium with uranium-spiked groundwater, the influent solution was changed to Hanford groundwater containing the polyphosphate amendment or calcium followed by the other respective solution. Aliquots of the effluent solutions were collected and monitored using ICP-MS and ICP-OES. Several uranium mineral phases will fluoresce under ultraviolet (UV) radiation. This property can be used to rapidly and efficiently evaluate the presence of uranium phases within large sedimentary matrices. A control column was run to confirm that no other fluorescent phases were present within the sediment or formed as a result of saturating the column with 1 ppm uranium, indicating that fluorescence spectroscopy could be used for qualitative evaluation of uranium-phosphate mineral phase formation. The solid-phase formations were evaluated by fluorescence spectroscopy using short wave UV radiation, 254 nm.

DYNAMIC SEQUESTRATION OF URANIUM

Column experiments were conducted to quantify the efficacy of an in situ hydroxylapatite barrier on the retardation and retention of uranium. Two borosilicate glass columns (length, L = 5 cm; radius, r = 1.25 cm; and bulk volume, V_b = 28.55 cm^3) were packed uniformly with the < 2.00 mm fraction of sediment from cores collected from the 300 Area. One of the columns had a 5-wt% barrier of hydroxylapatite placed within the middle of the sediment bed (figure 7). The columns were saturated with Hanford groundwater until stable water content was attained. The influent solution was changed to Hanford groundwater spiked with 10^{-6} M uranium. Syringe pumps were used to control the flow rate, 40 mL/day. Effluent solution was collected continuously and aliquots of the fluid sample were retained for both pH measurement and analysis of calcium and phosphorus concentrations using ICP-OES and uranium using ICP-MS.

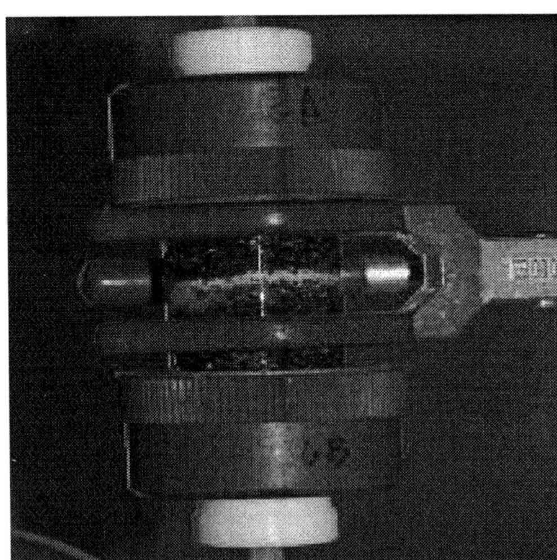

Figure 7. Photo Displaying the Apatite Barrier Placed in the Middle of a Sediment Column Composed of the < 2 mm Fraction from the 300 Area.

Table 3. Experimental Parameters for Polyphosphate Amendment Optimization

Column No.	Amendment Source	Wt% Phosphate Source	Injection Order	$Ca:P_{total}$	pH	Conc., M
1	Ortho $[P]_{aq}$	0.25	1	2.2	7	1.32×10^{-3}
	Pyro $[P]_{aq}$	0.25				6.58×10^{-4}
	Tripoly $[P]_{aq}$	0.5				8.77×10^{-4}
	Calcium		2			1.15×10^{-2}
2	Ortho $[P]_{aq}$	0.25	1	2.2	7	1.97×10^{-3}
	Pyro $[P]_{aq}$	0.25				9.87×10^{-4}
	Tripoly $[P]_{aq}$	0.5				1.32×10^{-3}
	Calcium		2			1.74×10^{-2}
3	Ortho $[P]_{aq}$	0.25	1	2.2	No adj.	1.97×10^{-3}
	Pyro $[P]_{aq}$	0.25				9.87×10^{-4}
	Tripoly $[P]_{aq}$	0.5				1.32×10^{-3}
	Calcium		2			1.74×10^{-2}
4	Ortho $[P]_{aq}$	0.375	1	2.2	No adj.	2.63×10^{-3}
	Pyro $[P]_{aq}$	0.25				1.32×10^{-3}
	Tripoly $[P]_{aq}$	0.375				1.75×10^{-3}
	Calcium		2			2.32×10^{-2}
5	Ortho $[P]_{aq}$	0.25	1	1.67	No adj.	3.47×10^{-3}
	Pyro $[P]_{aq}$	0.25				1.74×10^{-3}
	Tripoly $[P]_{aq}$	0.5				2.32×10^{-3}
	Calcium		2			2.32×10^{-2}
6	Ortho $[P]_{aq}$	0.25	1	1.67	7	3.47×10^{-3}
	Pyro $[P]_{aq}$	0.25				1.74×10^{-3}
	Tripoly $[P]_{aq}$	0.5				2.32×10^{-3}
	Calcium		2			2.32×10^{-2}

Table 3. (Continued).

Column No.	Amendment Source	Wt% Phosphate Source	Injection Order	Ca:P_{total}	pH	Conc., M
7/11	Ortho $[P]_{aq}$	0.25	1	2.2	No adj./7	2.63×10^{-3}
	Pyro $[P]_{aq}$	0.25				1.32×10^{-3}
	Tripoly $[P]_{aq}$	0.5				1.75×10^{-3}
	Calcium		2			2.32×10^{-2}
8/12	Ortho $[P]_{aq}$	0.25	1	2.2	No adj./7	6.58×10^{-3}
	Pyro $[P]_{aq}$	0.25				3.29×10^{-3}
	Tripoly $[P]_{aq}$	0.5				4.39×10^{-3}
	Calcium		2			5.79×10^{-2}
9/13	Ortho $[P]_{aq}$	0.25	1	2.2	No. Adj/7	9.21×10^{-3}
	Pyro $[P]_{aq}$	0.25				4.61×10^{-3}
	Tripoly $[P]_{aq}$	0.5				6.14×10^{-3}
	Calcium		2			8.10×10^{-2}
10/14	Ortho $[P]_{aq}$	0.25	1	2.2	No Adj./7	1.32×10^{-2}
	Pyro $[P]_{aq}$	0.25				6.58×10^{-3}
	Tripoly $[P]_{aq}$	0.5				8.77×10^{-3}
	Calcium		2			1.16×10^{-1}

Chapter 4

APATITE AND AUTUNITE LONGEVITY

In dynamic systems the long-term stability of minerals is controlled by the solubility and the dissolution rate of the mineral. Under highly advective conditions where transport is greater than the solubility, the stability of the mineral is controlled by dissolution kinetics. Alternatively, in low to moderately advective environments where solubility is greater than transport, the long-term stability of the mineral is based on the solubility of the phase. The former conditions are relevant to the 300 Area saturated zone. As such, to quantify longevity of autunite, apatite, and therein the performance of polyphosphate technology, it is necessary to determine the rate of autunite and apatite dissolution under conditions representative of the 300 Area aquifer.

SINGLE-PASS FLOW-THROUGH DISSOLUTION

Evaluation of the dissolution of apatite was performed with the single-pass flow-through (SPFT) test method. The SPFT apparatus provides for experimental flexibility allowing each of the kinetic test parameters to be isolated and quantified (McGrail and Peeler 1995; McGrail et al. 1997). Temperature, flow rate, solution composition, and sample mass and size can all be manipulated to assure accurate rate determinations (Wellman et al. 2006b, 2005c; Icenhower and Dove 2000; Icenhower et al. 2002, 2000, 2003; Pierce et al. 2005).

In general, the SPFT system (figure 8) consists of a programmable pump that transports solutions from an influent reservoir via Teflon lines. Solution is transferred into 60-mL capacity perfluoroalkoxide (PFA) reactors (Savillex). The reactors are situated within constant temperature ovens, whose temperature is controlled to ±2°C by tested and calibrated thermocouples. The powdered specimen rests at the bottom of the reactor and influent and effluent solutions enter and exit, respectively, from fluid transfer lines that protrude through two separate ports at the top of the reactor. The residence time of aqueous solutions in the reactor varies with the flow rate, which is adjusted in accordance to the needs of the experiment. The effluent line carries solution to collection vials that are positioned outside the oven.

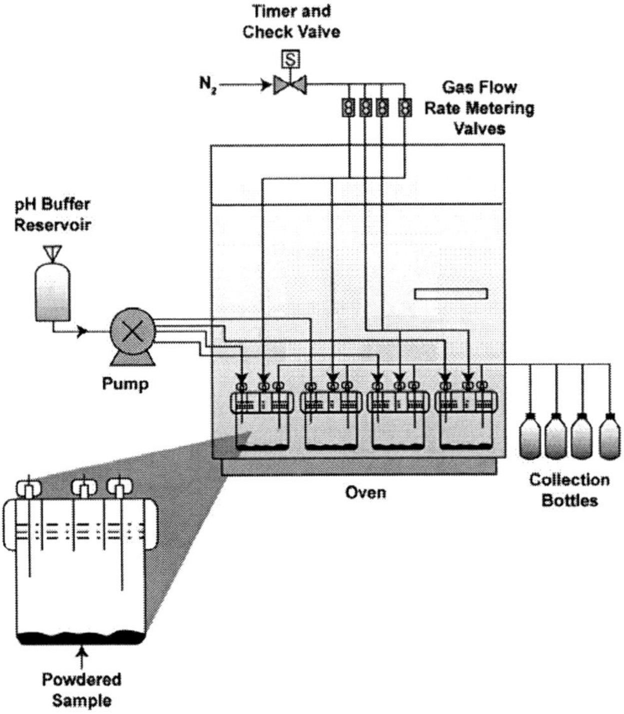

Figure 8. Schematic of the Single-Pass Flow-Through Dissolution Test System.

Effluent solution was collected continuously and aliquots of the fluid sample were retained for both pH measurement and analysis of dissolved element concentrations by ICP-OES and ICP-MS. Solutions earmarked for analysis by ICP-OES and ICP-MS were preserved in Optima™ nitric acid. Concentrations of aqueous matrix forming elements for apatite (calcium and phosphorus) and autunite (uranium and phosphorus) were used to quantify the dissolution rates as a function of pH and temperature. Before the sample specimens were added to the reactor, blank solution samples were collected and used to establish the concentration of background analytes. The blank samples were treated in exactly the same manner as the samples.

STARTING MATERIALS

The hydroxylapatite sample used in this study was obtained from Sigma-Aldrich Chemical and prepared by sieving into the desired size fractions with American Society for Testing and Materials (ASTM) standard sieves (ASTM 2002). The surface area was determined using N_2-adsorption Brunauer-Emmett-Teller (BET) analysis (Brunauer et al. 1938), 74.25 m^2/g.

Synthetic sodium meta-autunite I, $Na_2[(UO_2)(PO_4)]_2 \cdot 3H_2O$ (herein designated Na-autunite), was prepared by direct precipitation from a mixture of uranyl nitrate with sodium phosphate, dibasic. The precipitated phase was characterized using extended x-ray absorption fine structure (EXAFS) spectroscopy, chemical digestion followed by ICP-OES and ICP-MS for elemental analyses, x-ray diffraction (XRD), scanning electron microscopy (SEM), and multi-point BET analyses (Wellman et al. 2005a).

Natural calcium meta-autunite I, $Ca[(UO_2)(PO_4)]_2 \cdot 3H_2O$ (herein designated GHR) was obtained from northeastern Washington State. The material was characterized using ICP-OES and ICP-MS analyses, XRD, and SEM to confirm the composition, structure, and morphology of the autunite minerals as 98 to 99% pure autunite with calculated anhydrous structural formula consistent with Ca-autunite: $Ca[(UO_2)(PO_4)]_2$. Electron microprobe analyses further indicated that the autunite mineral contains ~3 waters of hydration per formula unit (p.f.u.). Powdered samples of Na-Autunite and Ca-autunite (GHR) were prepared to be within the same size fraction, 25 to 45 μm (-325 to +500 mesh); however, the surface cracking, fractures, and basal plane cleavage of the GHR resulted in a greater surface area relative to Na-autunite (table 4). Therefore, a second size fraction of GHR (75 to 150 μm, or -100 to + 200 mesh) was prepared that had a comparable surface area to Na-autunite (table 4).

Table 4. Source, Particle Size, and Surface Area of Autunite Minerals

Autunite Composition	Sample ID	Source	Particle Size	Surface Area, m^2/g
$Na_2[(UO_2)(PO_4)]_2 \cdot 3H_2O$	Na-Autunite	Synthetic	-325 +500	0.78
$Ca[(UO_2)(PO_4)]_2 \cdot 3H_2O$	GHR	Natural	-325 +500	2.30
$Ca[(UO_2)(PO_4)]_2 \cdot 3H_2O$	GHR	Natural	-100 +200	0.88

BUFFER SOLUTIONS

The pH values of the solutions used in these experiments spanned the range from 6 to 12. Maintaining a constant temperature and flow rate, while varying the solution pH, allows the value of the pH power law coefficient, η, to be determined. The solutions used to control the pH during the SPFT experiments are summarized in table 5, which also contains a summary of the in situ pH values computed at each test temperature using EQ3NR (Wolery 1992). It is important to take into account the change in pH that occurs at different temperatures when computing dissolution rates from SPFT data because the in situ pH can vary by as much as 1.5 pH units over the temperature range from 23° to 90°C. By quantifying temperature and pH-dependent rate parameters the dissolution rate of hydroxylapatite can be extrapolated to conditions representative of the subsurface. Buffer solutions were prepared by adding small amounts of the organic tris hydroxymethyl aminomethane (TRIS) buffer to deionized water (DIW) and adjusting the solution to the desired pH value using 15.8M HNO_3 or 1 M LiOH. The TRIS buffer range is between pH 7 to 10; therefore, the alkaline solutions, pH range 11 and 12, were prepared by adding of LiOH and LiCl to DIW and adjusting the solution to the desired pH value using 15.8 M HNO_3 or 1M LiOH.

Table 5. Composition of Solutions Used in Single-Pass Flow-Through Experiments. Solution pH values above 23°C were calculated using the EQ3NR Code V7.2b database

Solution	Composition	pH @			
		23°C	40°C	70°C	90°C
1	0.05 M TRIS + 0.047 M HNO_3	7.01	6.57	5.91	5.55
2	0.05 M TRIS + 0.02 M HNO_3	8.32	7.90	7.25	6.89
3	0.05 M TRIS + 0.0041 M HNO_3	8.99	8.67	8.08	7.72
4	0.05 M TRIS + 0.003 M LiOH	9.99	9.55	8.88	8.52
5	0.0107 M LiOH + 0.010 M LiCl	11.00	10.89	10.43	10.06
6	0.0207 M LiOH + 0.010 M LiCl	12.02	11.74	11.08	10.70

TRIS = Tris hydroxymethyl aminomethane buffer.

Rate Calculations and Uncertainty

Dissolution rates, based on steady-state concentrations of elements in the effluent solution, are normalized to the amount of the element present in the sample by the following formula:

$$r_i = \frac{\left(C_i - \bar{C}_{i,b}\right) q}{f_i\, S} \tag{1}$$

where

r_i = the normalized dissolution rate for element i (g m^{-2} d^{-1})
C_i = the concentration of the element i in the effluent (g L^{-1})
$\bar{C}_{i,b}$ = the average background concentration of the element of interest (g L^{-1})
q = the flow rate (L d^{-1})
f_i = the mass fraction of the element in the metal (dimensionless)
S = the surface area of the sample (m^2).

The value of f_i was calculated from the chemical composition of the sample. Flow rates are determined by gravimetric analysis of the fluid collected in each effluent collection vessel upon sampling. The background concentration of the element of interest is determined, as discussed previously, by analyses of the starting input solution and three blank solutions. Typically, background concentrations of elements are below their respective detection threshold. The detection threshold of any element is defined here as the lowest calibration standard that can be determined reproducibly during an analytical run within 10%. In cases where the analyte is below the detection threshold, the background concentration of the element is set at the value of the detection threshold.

Determining the experimental uncertainty of the dissolution rate takes into account uncertainties of each parameter in Equation (1). For uncorrelated random errors, the standard deviation of a function $f(x_1, x_2, \ldots x_n)$ is given by

$$\sigma_f = \sqrt{\sum_{i=1}^{n} \left(\frac{\partial f}{\partial x_i}\right)^2 \sigma_i^2} \tag{3}$$

where

σ_f = the standard deviation of the function f
x_i = parameter i
σ_I = the standard deviation of parameter i.

Substituting Equation (1) into (2) results in the following:

$$\sigma_{r_i} = \sqrt{\left(\frac{q}{f_i S}\right)^2 (\sigma_{C_i}^2 + \sigma_{\bar{C}_{i,b}}^2) + \left(\frac{C_i - \bar{C}_{i,b}}{f_i S}\right)^2 \sigma_q^2 + \left(\frac{(C_i - \bar{C}_{i,b})q}{f_i^2 S}\right)^2 \sigma_{f_i}^2 + \left(\frac{(C_i - \bar{C}_{i,b})q}{f_i S^2}\right)^2 \sigma_S^2} \quad (4)$$

Equation (2) can also be expressed in terms of the relative error, $\hat{\sigma}_{r_i} = \sigma_{r_i} / r_i$, and is given by

$$\hat{\sigma}_{r_i} = \sqrt{\frac{\left(\hat{\sigma}_{C_i} C_i\right)^2 + \left(\hat{\sigma}_{\bar{C}_{i,b}} \bar{C}_{i,b}\right)^2}{\left(C_i - \bar{C}_{i,b}\right)^2} + \hat{\sigma}_q^2 + \hat{\sigma}_{f_i}^2 + \hat{\sigma}_S^2} \quad (5)$$

Relative errors of 10%, 10%, 5%, 3%, and 15% for C_i, $\bar{C}_{i,b}$, q, f_i , and S, respectively, are typical for measurements conducted at Pacific Northwest National Laboratory. But to reduce the error associated with mass fraction (f_i), the samples used in these experiments were ground, homogenized, sub-sampled, and analyzed at least three times to obtain a more accurate composition with a better estimate of the uncertainty. The conservative appraisal of errors assigned to the parameters in Equation (4), in addition to the practice of imputing detection threshold values to background concentrations, results in typical uncertainties of approximately ±35% on the dissolution rate.

Chapter 5

MATERIALS AND METHODS FOR PILOT-SCALE FIELD DEMONSTRATION OF POLYPHOSPHATE TECHNOLOGY

Focused application of polyphosphate technology is proposed in source or "hot spot" areas that would significantly reduce the inventory of available uranium that contributes to the groundwater plume. Figure 9 presents a schematic depiction of the pilot-scale field demonstration.

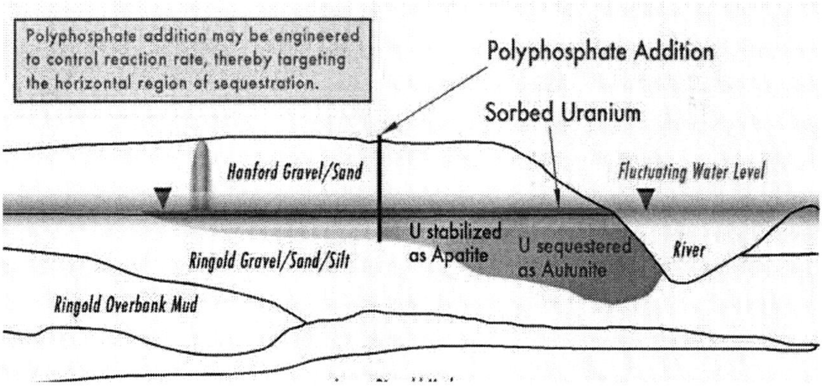

Figure 9. Schematic Depicting the 300-FF-5 Operable Unit Geology and Proposed Treatability Test of Polyphosphate to Sequester Uranium.

The polyphosphate amendment was injected into the uranium plume to immediately treat the aqueous uranium present through precipitation of autunite. Subsuequently, lateral formation of apatite within the sedimentary matrix serves as a long-term remediation scheme to sequester uranium further leached into the aquifer. The presence of apatite serves as a long-term sorbent of uranium and, also, as a source of phosphate for the continued formation of uranyl-phosphate minerals.

Test Site Location

The pilot-scale field test was conducted at a location in the Hanford 300 Area containing the highest aqueous uranium concentrations (>120 ppb) in the plume (figure 10). Well 399-1-23 was selected as the candidate well for performing the treatability test (figure 12) (Vermuel et al. 2006). This well is located at the southern end of a liquid waste disposal trench that is a known source of uranium. The well is sufficiently far inland from the Columbia River that infiltration of river water during normal high discharge conditions does not reach the well, which if that occurred, would complicate interpretation of test results. This well also was selected because of its contaminant profile; i.e., it has a moderately thick (~6.1 m [20 ft]) interval of contaminated permeable Hanford gravel below the water table, which is less impacted by river level fluctuations than locations closer to the river. Depth-discrete uranium concentrations in groundwater samples collected during drilling in this well ranged up to 202 μg/L in the uppermost portion of the aquifer and dropped off to below detection deep in the aquifer below the Hanford-Ringold contact (figure 11). The highest concentration observed in the profile is located just below the Hanford-Ringold contact, indicating that vertical gradients generated during liquid waste disposal activities at the 316-5 process trenches likely drove contamination into the uppermost portion of the Ringold Formation.

Although the relatively low permeability of the Ringold Formation and cessation of liquid discharge to the trenches results in limited potential for additional transport of uranium within the Ringold Formation, these contaminated sediments will continue to act as a deep uranium source at this location.

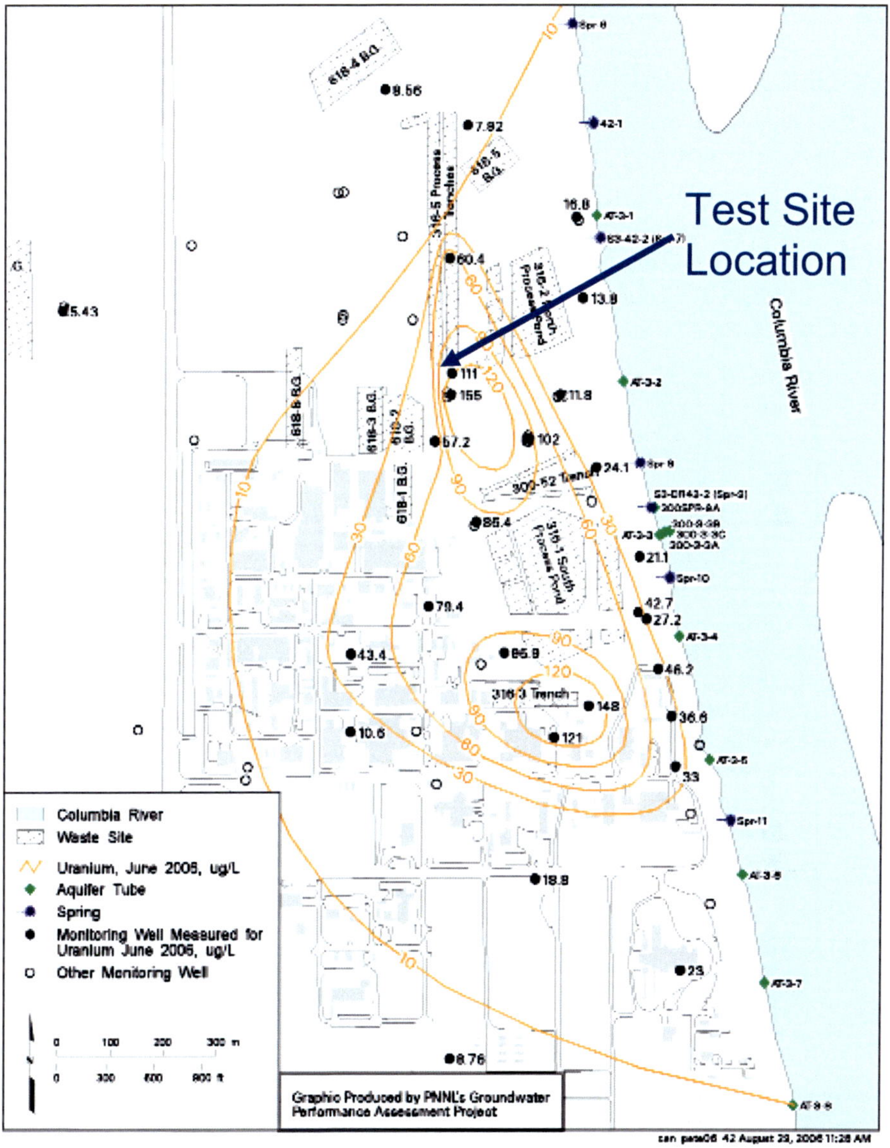

Figure 10. Contour Map of the 300 Area Uranium Plume and Location of Polyphosphate Pilot-Scale Test.

Tracer Test

A tracer injection test was conducted at the polyphosphate treatability test site on December 13, 2006. The objective of the tracer test was to further evaluate formation heterogeneities, assess the down-gradient transport of the tracer plume (i.e., aquifer transport properties), refine the polyphosphate injection design, and test operational procedures. Results from the tracer test provide information on the effective porosity of the aquifer, expected arrival times at the monitoring wells, and polyphosphate solution volume required for the targeted treatment zone thickness and radial extent.

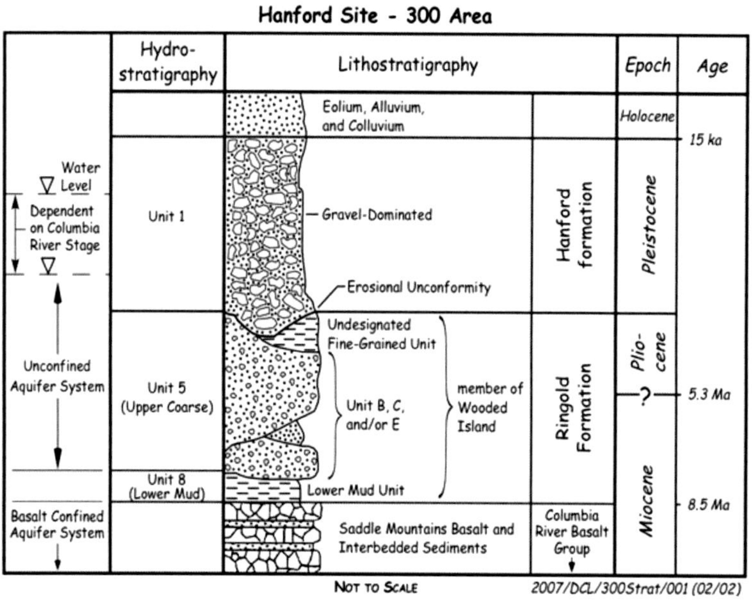

Figure 11. Hydrogeologic Column Depicting the Hydrogeology of the 300 Area (Williams et al. 2007).

The tracer test was conducted by injecting a solution containing a conservative, non-reactive bromide (Br) tracer into a central injection well (399-1-23, as shown in figure 12). Bromide concentrations were measured in the injection stream and the surrounding monitoring wells to determine the arrival times and extent of the tracer plume. Table 6 summarizes the operational parameters for the tracer test. The concentrated bromide solution was prepared in an ~151 L (40 gal) plastic drum and diluted in-line during the injection to the required concentration using withdrawn groundwater from Well 399-1-7, located ~188.9 m (620 ft) downgradient from the injection well. The concentrated bromide solution consisted of 60.8 kg (134 lb) NaBr mixed with 99.6 L (26.3 gal) of de-ionized water, for an approximate NaBr concentration of 610 g/L. The injection stream was maintained at a constant rate of 757 L (200 gal) per minute throughout the test duration. The concentrated solution was delivered to the injection stream at an average flow rate of 0.14 L/min (2.2 gal/hr). This resulted in an average injection concentration of ~112 mg/L NaBr, or 87 mg/L Br^-. The NaBr solution was injected into the aquifer through the injection well (399-1-23) for 11.9 hours (714 minutes), yielding a total injection volume of 541,300 L (143,000 gal). Flow rates for the injection stream during the test were monitored using in-line turbine flow meters and continuously recorded on a data logger.

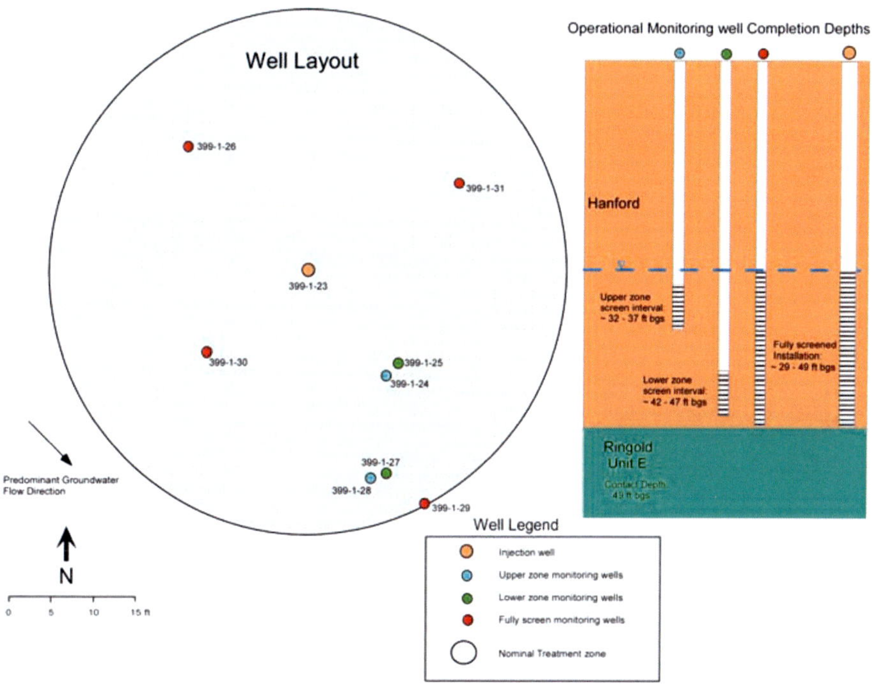

Figure 12. Polyphosphate Treatability Test Site Showing Well Layout.

Bromide concentrations were monitored in the injection stream and monitoring wells to determine the effected radial extent of the tracer plume during the test. Downhole ion-selective electrode (ISE) probes continuously monitored bromide concentrations in the wells during the test. A total of 256 aqueous samples were collected from the injection stream and surrounding monitoring wells and were analyzed in the field laboratory trailer for bromide using an ISE probe. Specific conductance (SpC), dissolved oxygen (DO), pH, oxidation-reduction potential (ORP), and temperature also were measured using an in-line electrode in the sampling manifold. The ion chromatography (IC) analyses were conducted on each of the 256 archive samples at an offsite laboratory as an additional method of measuring bromide concentration.

Table 6. Summary of Polyphosphate Treatability Test Site Tracer Injection Test

Test Parameter	Value
Tracer Mass	60.8 kg (134.0 lbs) of Sodium Bromide (NaBr)
Concentrated Tracer Solution Volume	26.3 gal (99.6 L)
Total Injection Rate	200 gal/min (757 L/min)
Conc. Tracer Injection Rate	2.2 gal/hr (0.14 L/min)
Make-up Water from 399-1-7 Injection Rate	200 gal/min
Injection Concentration	87 mg/L Br^-
Injection Duration	714 min (11.9 hr)
Injection Volume	142,600 gal

Unit Abbreviations: kg = kilogram; gal = gallon; L = liter; min = minute; gal/min = gallon per minute; L/min = liter per minute; gal/hr = gallon per hour; mg/L = milligrams per liter.

POLYPHOSPHATE PILOT-SCALE FIELD TEST DEMONSTRATION

The polyphosphate injection test was conducted June 11-15, 2007. Based on laboratory optimization on site-specific geologic media, a three-phase injection strategy was identified as an effective approach for obtaining both direct treatment of the uranium contamination in groundwater (i.e., autunite formation) and secondary formation of calcium-phosphate. This approach is designed to provide the long-term treatment capacity within the amended zone to address uranium solubilized and released from the deep vadose zone and capillary fringe during future high water table conditions. The three-part injection strategy consists of the following:

1. Initial polyphosphate amendment injection to precipitate aqueous uranium within the treatment zone as autunite. This prevents the formation of soluble calcium-uranate, which may redissolve, thereby

releasing a pulse of uranium into the groundwater upon injection of the soluble polyphosphate.

2. The initial polyphosphate injection was directly followed by injection of a calcium-chloride ($CaCl_2$) solution to provide a sufficient calcium source for apatite formation during a subsequent polyphosphate injection. Due to the higher partition coefficient (K_d) of the $CaCl_2$ solution as measured on site-specific sediments, a larger injection volume was required to reach the full radial extent of the targeted treatment zone for this component of the amendment formulation. However, this same increased retardation helped to facilitate mixing between the calcium and polyphosphate amendments during the third and final injection phase.
3. The $CaCl_2$ injection was directly followed by a final polyphosphate injection. This provided additional time-released phosphorus for lateral precipitation of calcium-phosphate as the remedy migrates downfield, and additional hydraulic driving force to achieve the maximum lateral distribution of solid-phase calcium-phosphate.

The concentrated injection solutions were prepared in a high-density polyethylene tanks and diluted in-line during the injection to the required concentration using withdrawn groundwater from Well 399-1-7, located ~188.9 m (620 ft) downgradient from the injection well (figure 13). The injection scheme consisted of three-phase, sequential injections of: 1) 250K gal initial injection of polyphosphate, 2) 500K gal of $CaCl_2$ solution, and 3) 250K gal final injection of polyphosphate to complete the treatment. The amendment formulations are listed in Table 7.

Figure 13. Field-Test Site Layout Centered Around the Injection and Monitoring Wells.

Table 7. Pilot-Scale Field Test Amendment Formulation

Injection	Amendment	Formula	Formula Wt, g/mol	Solubility, g/L 23°C H_2O	Density, g/cm^3 (25°C)	Conc., g/L	Conc., M
1	Sodium phosphate, monobasic	NaH_2PO_4	119.98	29.63	1.004	0.59	4.94×10^{-3}
	Sodium pyrophosphate	$Na_4P_2O_7$	265.9	32.81		0.66	2.47×10^{-3}
	Sodium tripolyphosphate	$Na_5P_3O_{10}$	367.86	60.40		1.21	3.29×10^{-3}
	Sodium bromide	NaBr	102.90			0.103	1.00×10^{-3}
2	Calcium chloride	$CaCl_2$	110.98	800	1.005	3.41	3.07×10^{-2}
3	Sodium phosphate, monobasic	NaH_2PO_4	119.98	29.63	1.004	0.59	4.94×10^{-3}
	Sodium pyrophosphate	$Na_4P_2O_7$	265.9	32.81		0.66	2.47×10^{-3}
	Sodium tripolyphosphate	$Na_5P_3O_{10}$	367.86	60.40		1.21	3.29×10^{-3}
	Sodium bromide	NaBr	102.90			0.103	1.00×10^{-3}

For tracing purposes, NaBr also was injected during phase 1 and 3 amendments; chloride was used as the tracer during the second phase injection. The injection stream was maintained at a constant rate of 757 L (200 gal) per minute throughout the test duration. All solutions were injected at a rate of 200 gpm. Flow rates for the injection stream during the test were monitored using in-line turbine flow meters and continuously recorded on a data logger.

Bromide or chloride concentrations were measured in the injection stream and the surrounding monitoring wells to determine the arrival times and extent of the tracer plume. Downhole ISE probes continuously monitored bromide or chloride concentrations in the wells during the test. One thousand aqueous samples were collected from the injection stream and surrounding monitoring wells and were analyzed in the field laboratory trailer for bromide and chloride using an ISE probe. Specific conductance (SpC), dissolved oxygen (DO), pH, oxidation-reduction potential (ORP), and temperature were also measured using an in-line electrode in the sampling manifold. The 1000 archived samples were analyzed at an offsite laboratory using ICP-MS and ICP-OES for complete cation analysis and IC for phosphate, sulfate, nitrate, nitrite, and as an additional method of measuring bromide and chloride concentrations. Aqueous performance monitoring samples also were collected one week, five weeks, and two months following treatment. The samples were subjected to comparable analytical analyses.

Chapter 6

RESULTS AND DISCUSSION

POLYPHOSPHATE HYDROLYSIS

A long-chain polyphosphate molecule is required to forestall the hydrolysis reaction of polyphosphate molecules and release of the orthophosphate molecule (PO_4^{3-}). However, a balance between the rate of polyphosphate degradation, groundwater flow rate, autunite and/or apatite precipitation, and injection rate must be met to optimize the remediation strategy. Thus, a clear understanding of polyphosphate hydrolysis kinetics is necessary to select the best chain, or mix of polyphosphate chain lengths, to directly precipitate autunite for immediate mitigation of aqueous uranium concentrations, and further precipitate apatite to control the long-term release of uranium from the sedimentary source.

In a homogeneous environment, the release of PO43- is dependent upon both the chain length and the pH of the solution; as the length of the phosphate chain increases, the hydrolysis rate decreases (Shen and Morgan 1973). Aqueous cations are believed to accelerate the hydrolysis reaction for tripolyphosphate by withdrawing the electron density from around the central phosphorus atom, thereby allowing nucleophilic attack by water molecules (Kura 1987; Kura and Tsukuda 1993; Shen and Dyroff 1966; Watanabe et al. 1975; Wazer and Griffith 1955). However, surface-mediated processes affect reaction rates in heterogeneous systems by lowering the activation energy, Ea, of the system. Therefore, it is essential to quantify the hydrolysis rates of long-chain phosphates in porous media before a remediation strategy can be implemented effectively. Although significant research has been conducted on cation-accelerated processes, experimental conditions applicable to the Hanford saturated zone have not been examined. Controlled application of polyphosphate technology within the subsurface required understanding the effect of aqueous cations, pure minerals, and native Hanford Site sediment on the hydrolysis of tripolyphosphate.

Figure 14 shows the ^{31}P NMR spectra of 0.2 M tripolyphosphate and 0.2 M pyrophosphate in carbonate buffered D_2O. Tripolyphosphate spectra show three distinct signals at ~-3 ppm, -4.7 ppm, and ~-18 ppm. The resonant signal at -3 ppm (doublet) and -18 ppm (triplet) represent tripolyphosphate, whereas the single peak observed at -4.7 ppm represents the degradation reaction product pyrophosphate. Tripolyphosphate degrades to pyro- and orthophosphate as shown in Equation (5); once formed pyrophosphate can undergo further hydrolysis to orthophosphate, Equation (6).

$$\underbrace{P-O-P-O-P}_{Tripoly} \rightarrow \underbrace{P-O-P}_{Pyro} + \underbrace{P}_{Ortho} \tag{5}$$

$$\underbrace{P-O-P}_{Pyro} \rightarrow \underbrace{P}_{Ortho} + \underbrace{P}_{Ortho} \tag{6}$$

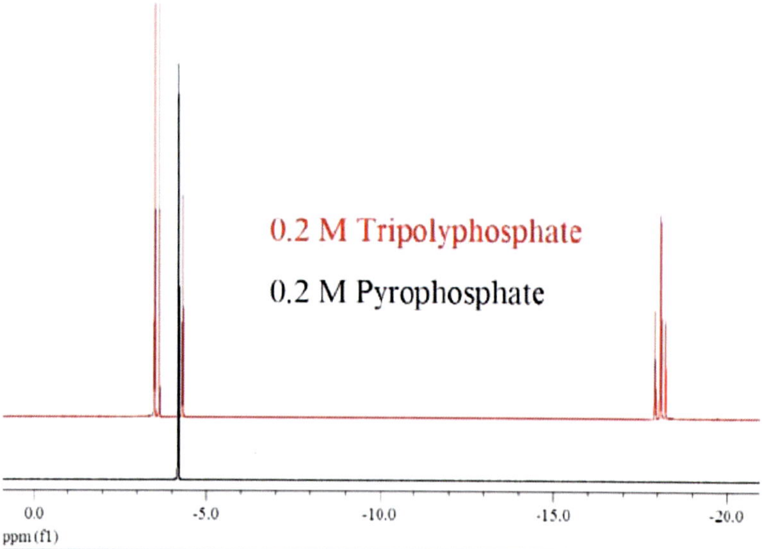

Figure 14. A ^{31}P NMR Spectrum of a Buffered Aqueous Solution of 0.2 M Pyro- and Tripolyphosphate Solutions. A single peak is displayed in the pyrophosphate spectra at -4.2 ppm, whereas the tripolyphosphate spectra show three signals: 1) the tripolyphosphate triplet (~ -18 ppm), 2) the tripolyphosphate doublet (~ -3 ppm), and 3) the pyrophosphate degradation compound (~ -4.7 ppm).

Prior to conducting the homogeneous and heterogeneous degradation experiments, a series of ^{31}P NMR experiments were conducted with known amounts of tripolyphosphate. The results from these experiments provided the required information needed to develop a linear relationship between the concentration of tripolyphosphate and integrated peak area (figure 15). Equation (7) is based on the spectra obtained for the tripolyphosphate doublet and the results of a linear regression are shown in figure 15.

$$P\,Conc. = \frac{Integrated\ Peak\ Area}{\left(2.94\times10^{6}\right)} \qquad (7)$$

The resulting regression coefficient is $(2.94 \pm 0.20) \times 10^6$ with a $R^2 = 0.99$. A similar technique was used for developing equations to quantify the degradation products, pyro- and orthophosphate.

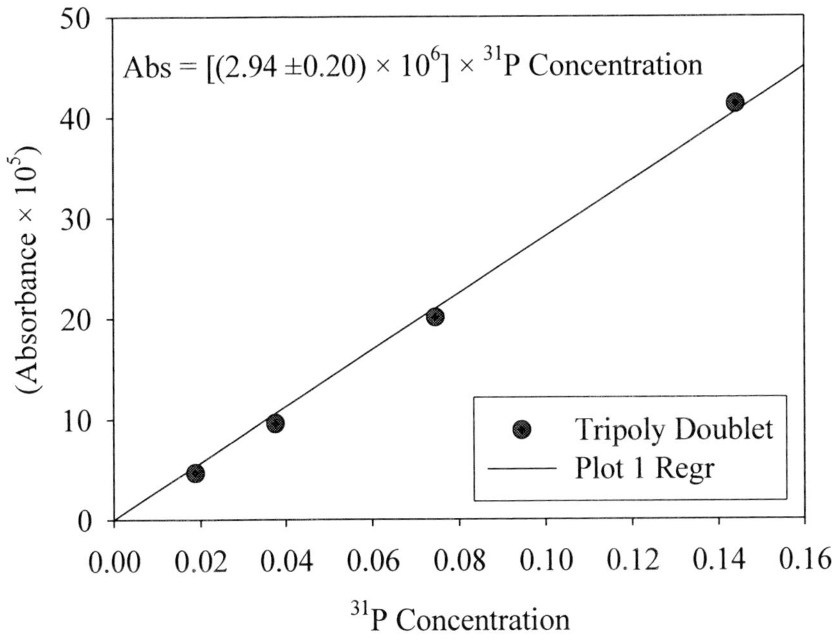

Figure 15. A 31P NMR Spectrum Integrated Peak Area as a Function of a Known Aqueous Concentration of Tripolyphosphate. These results are based on the tripolyphosphate doublet and were used to develop a linear equation that could be used to quantify the amount of tripolyphosphate at a given time.

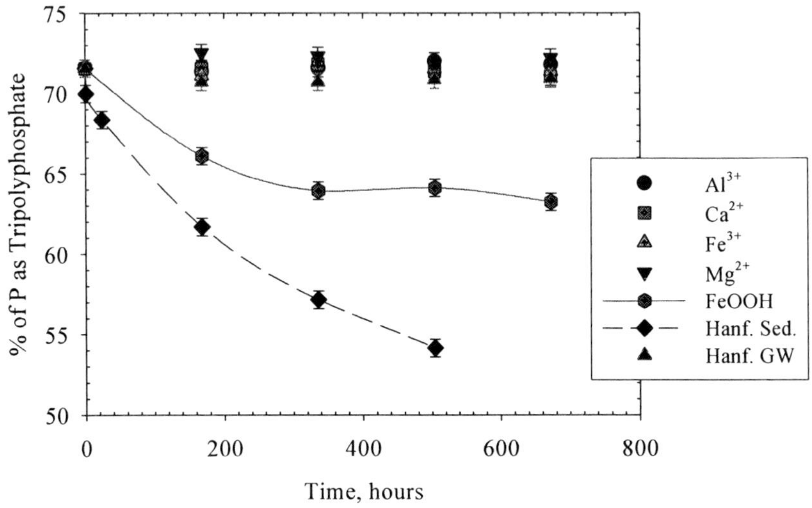

Figure 16. Percentage of Phosphorus as Tripolyphosphate as a Function of Time for Homogeneous Experiments Conducted with Aqueous Cations, Al^{3+}, Ca^{2+}, Fe^{3+}, Mg^{2+}, and Hanford Site Groundwater; and Heterogeneous Experiments Conducted with Solids, FeOOH, and Hanford Site Sediment.

Results from homogeneous ^{31}P NMR experiments suggest the presence of aqueous cations; Al^{3+}, Ca^{2+}, Fe^{3+}, and Mg^{2+} do not have any significant effect on the rate of tripolyphosphate hydrolysis at the cation concentrations used for these experiments (figure 16). These results are consistent with the findings of an independent homogeneous experiment conducted with Hanford Site groundwater, where the groundwater had no catalytic effect on tripolyphosphate degradation (figure 16). Also shown in figure 16 are the results collected for the heterogeneous experiments conducted with naturally occurring mineral FeOOH, as well as with native Hanford sediments, respectively. The only statistically significant deviations in the phosphate percentages were for the heterogeneous experiments. These results suggest the presence of FeOOH and native Hanford Site sediment had a measurable catalytic effect on the hydrolysis of sodium tripolyphosphate, as evidenced by 12% and 24% decreases in the tripolyphosphate concentration, respectively.

Chapter 7

APATITE AND AUTUNITE FORMATION

The following discussion summarizes the significant body of work within the literature regarding the formation of apatite from short-chain phosphate/polyphosphate aqueous solutions and underscores the importance of conducting site-specific testing to optimize polyphosphate technology for deployment in subsurface environments. Precipitation of apatite from homogeneous matrices has been suggested to proceed through initial precipitation of amorphous calcium-phosphate, which serves as a template for the heterogeneous nucleation of octacalcium-phosphate (OCP) (Feenstra and de Bruyn 1979). In turn, OCP serves as a template for epitaxial growth of hydroxylapatite (Feenstra and de Bruyn 1979; Brown et al. 1962; Eanes et al. 1965; Eanes and Posner 1965; Eanes and Meyer 1977). The conversion of amorphous to crystalline phases involving an epitaxial matching of the depositing phase onto the hydroxylapatite crystalline substrate is consistent with a hypothesized autocatalytic conversion mechanism (Boskey and Posner 1973; 1976; Eanes and Posner 1965); this explains the significance of apatite seed crystals for accelerated precipitation of hydroxylapatite from solution (Boskey and Posner 1973; Nancollas and Mohan 1970; Nancollas and Tomazic 1974; Inskeep and Silvertooth 1988; Brown 1981a, 1980, 1981b; Amjad et al. 1981). Once the reservoir of non-apatitic calcium-phosphate is depleted during the conversion process, the increase in size of apatite crystals proceeds by Ostwald ripening, in which the overall number of apatite crystals is reduced by consolidation and recrystallization (Eanes and Posner 1970). The Gibbs-Kelvin effect states the thermodynamic driving force for this mechanism is that the equilibrium solubility of smaller particles decreases with increasing particle size. Therefore, in a suspension of heterogeneously sized particles, the smaller particles have a higher solubility than larger particles, which causes the smaller particles to dissolve and the larger particles continue to grow (Eanes et al. 1965; Eanes and

Posner 1970). However, the growth rate of apatite is controlled by surface nucleation and/or dislocation mechanisms (Eanes and Posner 1970). As such, hydroxylapatite growth is limited by processes that occur at the crystal interface (Nancollas and Mohan 1970) and, therefore, are dependent on the surface area (Inskeep and Silvertooth 1988). Christoffersen and Christoffersen (1982) proposed that protonation of phosphate groups at the crystal surface catalyzes the exchange of phosphate between the apatite surface and the bulk solution, thereby accelerating growth. At pH 7.4, hydroxylapatite is the least soluble phase and most thermodynamically stable, in the absence of kinetic complications (Nancollas and Tomazic 1974). This is consistent with findings regarding the growth of fluorapatite (FAP), wherein a direct relationship between the growth rate of FAP and pH was observed (van Cappellen and Berner 1991). For a given degree of supersaturation, the growth rate of FAP at pH 7 was twice that measured at pH 8.

This underscores the complex series of elementary reactions in the precipitation of hydroxylapatite, which suggests either 1) direct precipitation from solution on the surface of hydroxylapatite seed crystals, or 2) precipitation from surface or absorbed calcium and phosphate whose concentrations are dependent on the solution calcium-to-phosphate ratio (Inskeep and Silvertooth 1988). The compactness of the heterogeneous nucleus is more conducive to the formation of hydroxylapatite than the diffuse homogeneous ionic nucleus (Garten and Head 1966). However, macromolecules can influence both the initial formation of amorphous calcium-phosphate and the conversion to apatite (Termine et al. 1970; Termine and Posner 1970). These macromolecules contain sites within their internal or solvation-shell that favors nucleation and growth (Termine et al. 1970; Termine and Posner 1970). Additionally, a decreased dielectric constant enhances initial mineral phase separation and amorphous-crystalline conversion. Thus, a partially apolar region within a macromolecule, as well as more polar regions, may provide a local milieu favorable for amorphous calcium-phosphate formation or crystal conversion (Termine et al. 1970). Sodium tripolyphosphate serves as a favorable nucleating surface for initial mineral phase separation and formation of amorphous calcium-phosphate with orthophosphate. When mineralization nucleation is considered relative to the initial mineral phase depositions, pyrophosphate will serve as a strong nucleating agent (Termine and Posner 1970).

Schmid and McKinney (1969) identified key processes involved in the formation of apatite from mixtures of ortho-, pyro-, and tripolyphosphate. Results of sorption studies illustrated that orthophosphate sorbs onto polyphosphate near pH ~7 to 9. Although, tripolyphosphate does not readily precipitate in the absence of orthophosphate, sorption of orthophosphate onto tripolyphosphate serves as a heterogeneous nucleating surface to promote precipitation. As orthophosphate begins to precipitate the pH of the solution increases slightly, while the degradation of tripolyphosphate is accelerated to form ortho- and pyrophosphate. This further enhances precipitation by providing additional orthophosphate. Furthermore, pyrophosphate produces a heavy, fast-settling precipitate with calcium, which increases the settling rate of the finer precipitates formed from tripolyphosphate. In the absence of orthophosphate, precipitation from tripolyphosphate is only ~50% of that under the same conditions in the presence of both ortho- and tripolyphosphate.

A key additional consideration regarding the use of a polyphosphate amendment in the precipitation of calcium-phosphate under conditions present within the 300 Area is the effect of carbonate. Precipitation of calcium-phosphate from monophosphate solutions is strongly influenced by competing reactions to produce calcium-carbonates (Lindsay and Moreno 1960; Diaz et al. 1994). Jenkins et al. (1971) demonstrated that in the $Ca\text{-}PO_4\text{-}CO_3\text{-}H^+\text{-}H_2O$ system, precipitation of calcium-carbonate competes with the precipitation of calcium-phosphate under the pH range of 9 to 10.5. Between pH 7.5 to 8.5 and above pH 10.5, calcium-phosphate precipitation controls the phosphorus concentration. For example, precipitation of calcium-phosphate at pH 8 initiated with an induction period of a couple hours followed by a period of rapid precipitation and prolonged slow removal of phosphorus from solution. As the bicarbonate concentration increased, the initial induction period required for precipitation of calcium-phosphate increased and the subsequent rate of removal as a function of bicarbonate concentration decreased.

By complexing calcium and sorbing to mineral surfaces, polyphosphate compounds effectively reduce both the rate and extent of calcium-carbonate precipitation, while simultaneously enhancing the rate of calcium-phosphate precipitation by reducing the competing reaction and "directing" calcium to participate in reactions resulting in calcium-phosphate precipitation (Ferguson et al. 1973).

Batch Experiments – Amendment Formulation and Efficacy

Initial batch tests were conducted based on the minimum amendment concentration as defined by previously conducted, preliminary column tests, which indicated a 1000 ppm sodium tripolyphosphate solution would reduce the aqueous concentration of uranium to near the MCL in ~12 pore volumes (Wellman et al. 2007b). This established the initial upper limit for the concentration of phosphorus at 1000 ppm. Additionally, lower concentrations of 100, 250, and 500 ppm were investigated to ensure that the amendment did not contain excessive phosphorus that may not be used in remediation efforts. Results of batch tests suggested that a concentration of 1000 ppm was required to remove more than 50% of the aqueous uranium.

Solubility limitations for calcium-phosphate sources prevented calcium dihydrogen phosphate, calcium hydrogen phosphate, and calcium pyrophosphate from serving as a sufficient source of phosphate or calcium to be included in the amendment formulation. Although, calcium hypophosphite provides a sufficient source of calcium and phosphorus, rather than forming discrete precipitates it produced fine floccules. The formation of fine floccules as a result of phytic acid remediation has been previously shown to provide sorption sites for uranium (Nash 2000; Nash et al. 1998a, 1997, 1998b, 1999). However, in open-framework, highly conductive subsurface environments fine floccules may be highly mobile. Alternatively, it has also been shown that rapid flocculation, due to heterogeneous nucleation, in regions of moderate to low hydraulic conductivity may occlude pore space (Wellman et al. 2006c). Either of these results would be detrimental and, therefore, calcium hypophosphite was eliminated from further consideration. Similar effects and concerns eliminated sodium trimetaphosphate from further consideration.

The data from the batch experiments are shown in figure 17. Figure 17 displays the percent of calcium and phosphorus removed from solution as a function of the calcium to phosphorus molar ratio in the presence of 10 and 1000 ppb uranium. The objective of these tests was to identify the calcium-to-phosphorus molar ratio for maximum removal from the aqueous phase. The mechanisms of removal may include sorption and/or precipitation; however, no attempt was made to discern the degree of removal based on these respective mechanisms. Greater than 90% removal of calcium and phosphorus from solution was achieved in the presence of sodium orthophosphate, sodium pyrophosphate, and sodium tripolyphosphate, respectively, with calcium-chloride as the calcium source (figure 17). The optimum molar ratio of calcium to phosphorus for sodium orthophosphate and sodium pyrophosphate is 1.5; whereas, the optimum calcium-to-phosphorus molar ratio for sodium tripolyphosphate is ~2.4. Moreover, removal of uranium from all solutions was both rapid (<2 min) and complete (~100%) indicating the formation of limited solubility compounds such as autunite and/or uranium-bearing apatite.

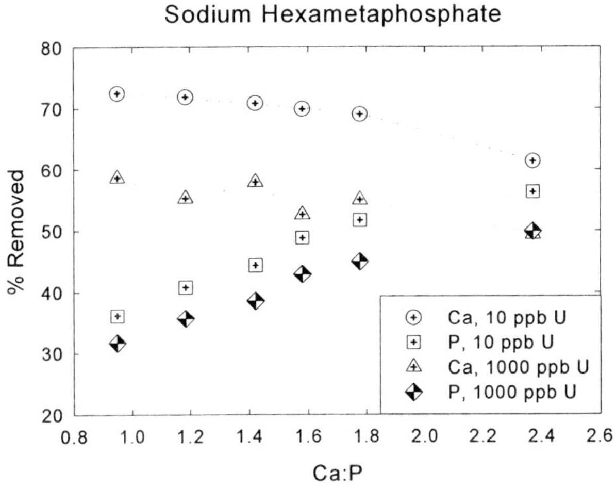

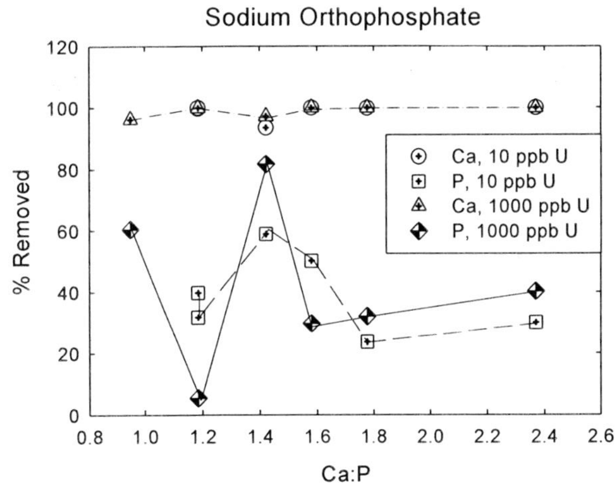

Figure 17. Continued on next page.

Sodium hexametaphosphate was eliminated from further consideration due to lower precipitation percentages and effects on the sedimentary matrix. These data suggested that it is possible to remove and sequester uranium from 300 Area aquifer using linear polyphosphate and calcium solutions.

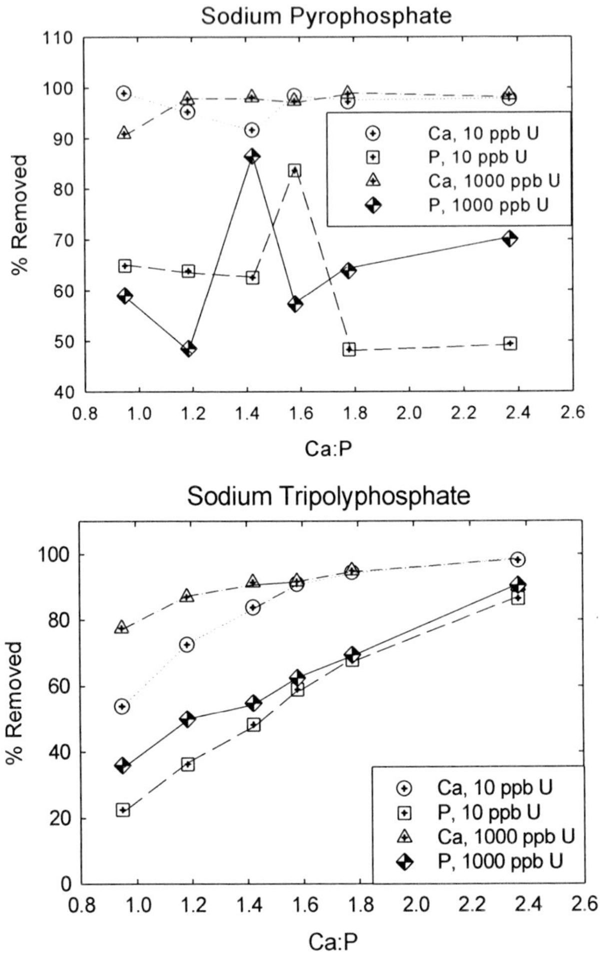

Figure 17. Percent Removal of Calcium and Phosphorus as a Function of Calcium-to-Phosphorus Molar Ratio.

Column Testing - Amendment Formulation, Efficacy, and Emplacement

Visual inspection of sediment removed from columns 1 through 4 (table 3), after application of the associated amendment formulations, indicated the formation of fluorescent green precipitates under shortwave (254-nm) UV radiation. A control column demonstrated that no phases were present or formed within the sedimentary matrix upon application of the polyphosphate amendment in the absence of uranium. Thus, the formation of fluorescent green precipitates was used as a means of qualitative identification of uranium-phosphate phases (figure 18). Qualitatively, the precipitate appeared to be within or coating ~50% of the sediment particles. Analysis of effluent solution samples by ICP-MS from columns 1 through 4 demonstrated ~50% reduction in the aqueous uranium concentration.

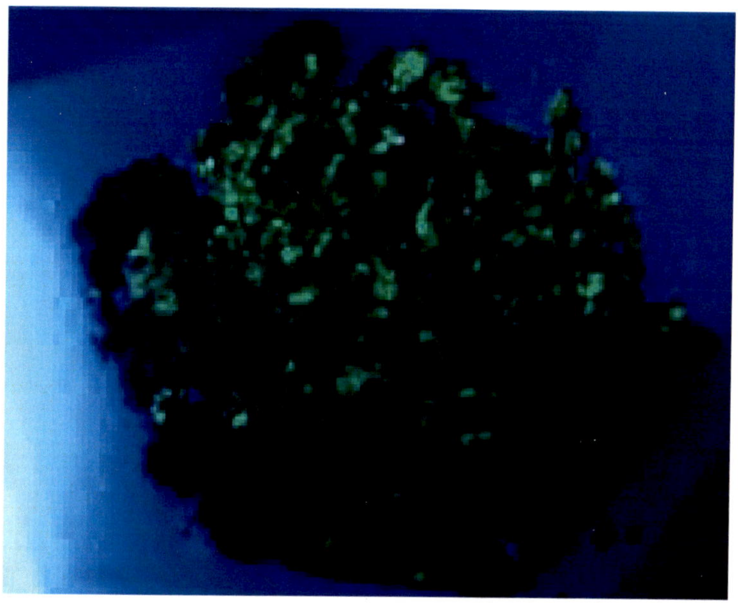

Figure 18. Representative Photo of Sediment Sectioned from the Effluent End of Column 1 Illustrating the Visual Identification of Uranium-Phosphate under Shortwave UV Radiation.

This suggested that to treat 100% of the aqueous uranium a higher concentration of phosphorus and calcium in the amendment formulation was necessary. Comparison of columns 2 and 3 suggested there was little effect of pH in reducing the aqueous uranium concentration; however, precipitation of calcium-phosphate was more significant under pH conditions ~7.

Column 4 highlighted the significance of the complex relationship between ortho-, pyro-, and tripolyphosphate. Although the concentration of aqueous uranium decreased ~50%, the formation of calcium-phosphate was restricted to a discrete region within the sediment matrix (figure 19, bottom).

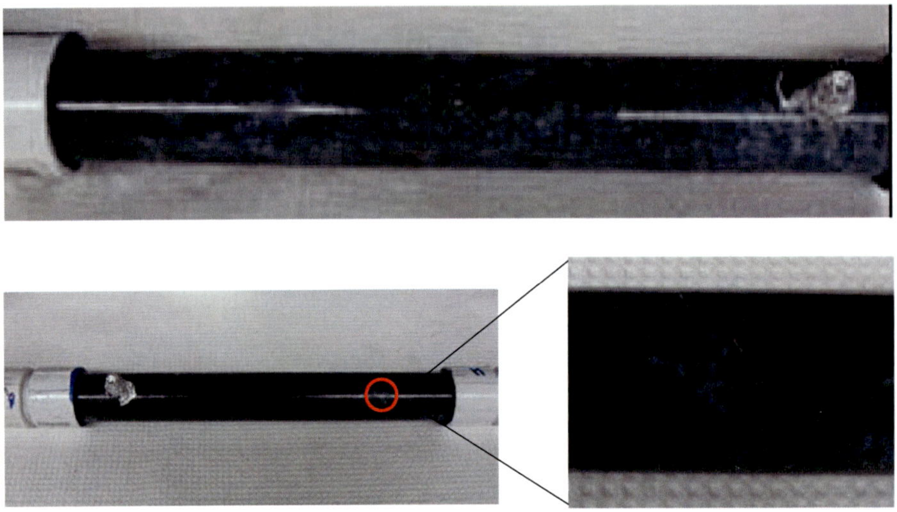

Figure 19. Photo Showing Disperse Precipitation of Calcium-Phosphate Throughout Column 1 (top) and Discrete Precipitation of Calcium-Phosphate Within Column 4 (bottom).

Columns 5 and 6 (Ca:P molar ratio = 1.67) in comparison to columns 2 and 3 (Ca:P molar ratio = 2.2) illustrated the significance of the calcium-to-phosphorus molar ratio. Qualitatively, the Ca:P molar ratio of 2.2 (columns 2 and 3) afforded more precipitation than a Ca:P molar ratio of 1.67 (columns 5 and 6), which gave no visual indication of calcium-phosphate precipitation. Although batch testing indicated the optimal Ca:P molar ratio for removal of calcium and phosphorus in the presence of both ortho-, and pyrophosphate was ~1.5, columns 1 through 4 illustrate the significance of the Ca:P ratio of 2.4 indicated by tripolyphosphate batch testing.

The calcium and phosphorus formulations were conducted in duplicate with columns 7 through 10 at the unadjusted pH (pH 7), and columns 11 through 14 at the adjusted pH with a Ca:P molar ratio = 2.2 for all columns. For these experiments, the concentration of calcium varied from 2.32×10^{-2} M to 1.16×10^{-1} M, and phosphorus concentrations ranged from 1.05×10^{-2} M to 5.26×10^{-2} M. Precipitation of calcium-phosphate in columns 7 through 10 was limited, eliminating consideration of non-adjusted amendment solutions. Alternatively, the degree of calcium-phosphate precipitation increased using the same amendment formulation adjusted to pH ~7, columns 11 through 14. In columns 11 and 12, the concentration of aqueous uranium in the effluent solution increased over the first 0.5 to 1 pore volume during remedy injection to concentrations between 1.2 to 3 times the influent uranium concentration (figure 20a, b). However, increasing the concentration of phosphorus and calcium in the amendment formulation (column 14) precluded this phenomenon. Additionally, the concentration of aqueous uranium in column 14 was reduced to below the MCL, 30 μg/L, within 0.5 to 1 pore volumes of treatment, and remained well below 30 μg/L for the remainder of the experiment (figure 20d).

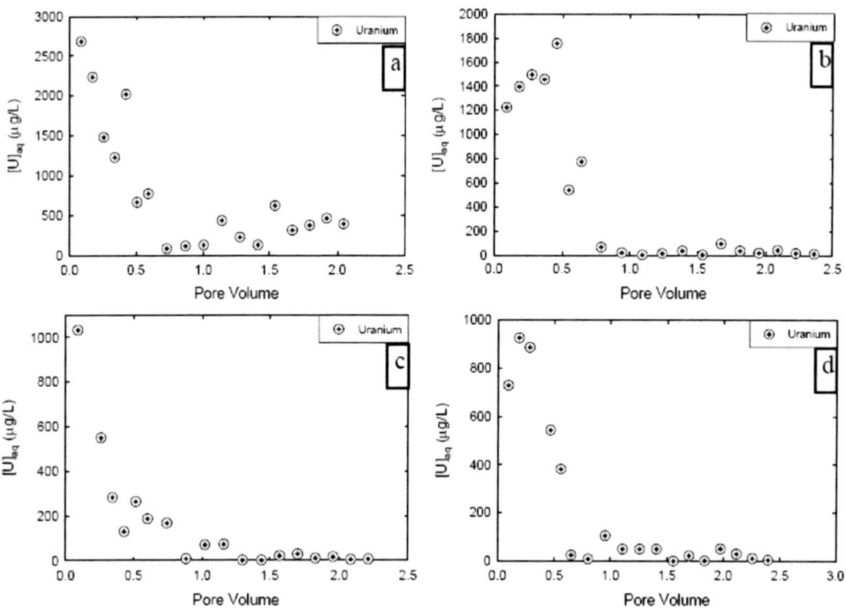

Figure 20. Graphs Depicting Aqueous Uranium Concentrations from Columns Saturated with 1,000 μg/L Uranium as a Function of the Number of Pore Volumes of Polyphosphate Remedy Displaced Through Columns a) 11, b) 12, c) 13, and d) 14 (table 3).

Columns 15 through 18 used the optimum formulations identified through previous tests (columns 13 and 14) as well as two additional formulations that contained equivalent total phosphorus concentrations, but maintained total calcium-to-phosphorus ratios of 1.9 (columns 17 and 18). The order of injection was altered for all columns (15 through 18) such that calcium was injected prior to phosphorus. Qualitative visual inspection of the columns following treatment suggests the most complete distribution within the column and removal of uranium occurred in column 16, using a calcium-to-phosphorus molar ratio of 1.9, pH 7 (figure 21).

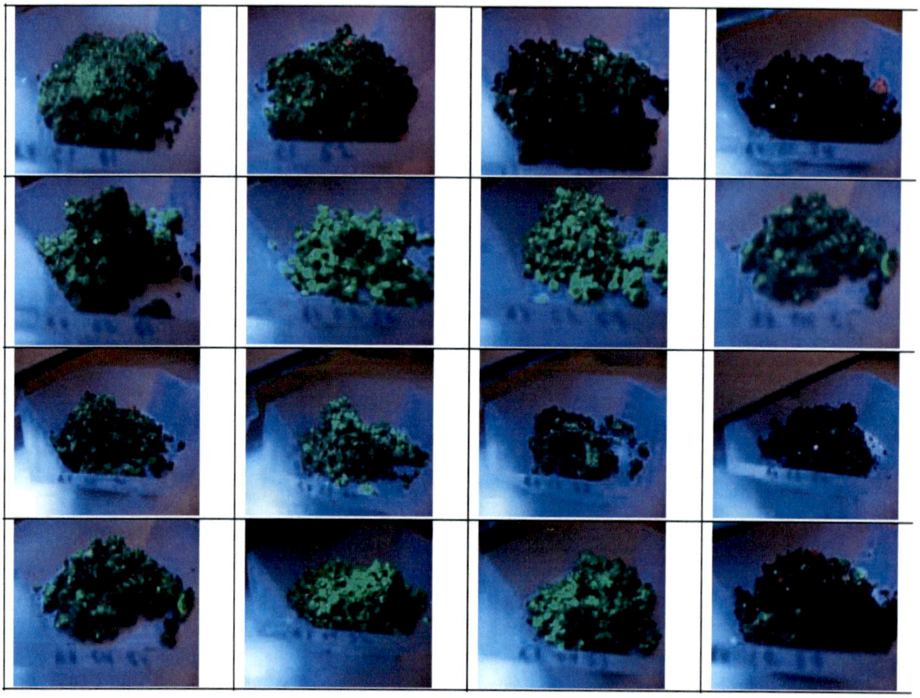

Figure 21. Photos of Column Sections Taken under Shortwave UV Radiation. Orientation: top-down, columns 15 through 18; left to right, influent to effluent.

However, with the exception of column 17, quantitative analysis of effluent uranium concentrations indicates that concentrations do not decline as rapidly as those measured in the previous set of columns, 11 through 14, wherein phosphorus was injected first followed by calcium (figure 22).

Additionally, the efficacy and long-term performance of columns 15 through 18 is less than that of columns 11 through 14 where uranium concentrations remained well below 30 μg/L. The aqueous concentration of uranium measured in the effluent solutions collected from columns 15 through 18 declined to below 30 μg/L, but then exhibited a number of fluctuations above and below the MCL for the remainder of the experiment. It is hypothesized that these fluctuations can be attributed to the initial formation of precursor calcium-uranate phases, which are more soluble than uranium-phosphate phases. Upon injection of the polyphosphate solution, the calcium-uranate phases likely undergo rapid dissolution to release soluble uranium that re-precipitates as a uranium-phosphate phase. Although both injection schemes ultimately result in formation of uranium-phosphate, precipitation and dissolution of calcium-uranate phases may afford undesirable fluctuations in uranium concentration above 30 μg/L.

Regardless of the injection order or concentration of phosphorus and calcium used in the amendment formulation, all phosphorus, including degradation products, was removed via sorption and/or precipitation reactions. Figure 23 is a representative plot for the removal of phosphorus during treatment of a uranium-contaminated column with results being comparable for all column tests conducted. Effluent concentrations of phosphorus are at or below background groundwater concentrations. Thus, the potential for downgradient transport and potential migration to the river is minimal.

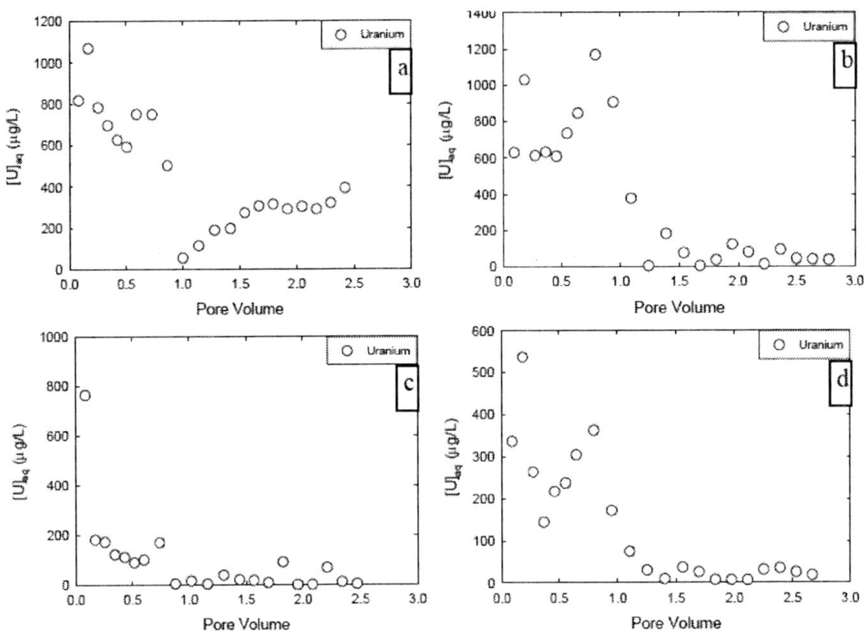

Figure 22. Graphs Depicting Aqueous Uranium Concentrations from Columns Saturated with 1000 μg/L Uranium as a Function of the Number of Pore Volumes of Polyphosphate Remedy Displaced Through Columns a) 15, b) 16, c) 17, and d) 18 (table 3). Remedy injection order was calcium followed by phosphorus.

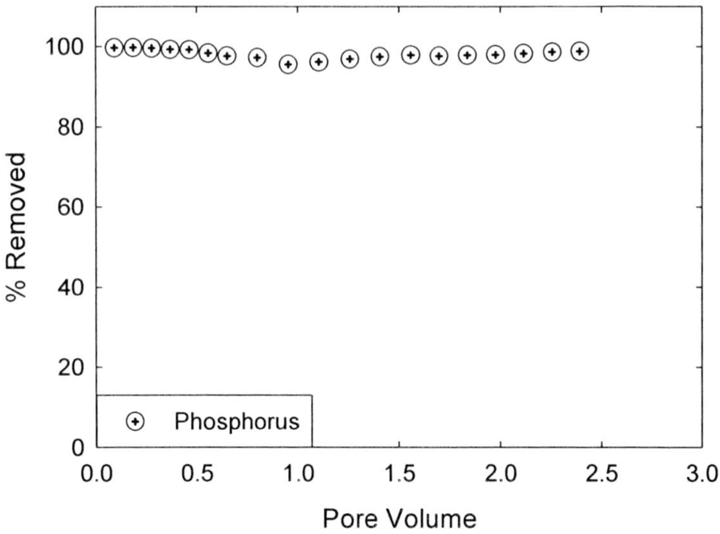

Figure 23. Representative Plot Depicting the Removal of Phosphorus by Sorption and Precipitation Reactions.

Chapter 8

DYNAMIC SEQUESTRATION OF URANIUM

Column experiments were conducted to quantify the efficacy of an in situ hydroxylapatite barrier on the retardation and retention of uranium. Sorption of uranium during transport, or retardation, R_{ef}, was calculated from the experimental breakthrough curve (BTC) using moment analysis, which is based on the center-of-mass and area under the BTC (Valocchi 1984, 1985). Apparent K_d values, K_{d-ap}, were calculated from R_{ef}, as follows

$$K_{d-ap} = (R_{ef} - 1)\frac{\theta}{\rho_b} \tag{8}$$

where θ is the volumetric water content and ρ_b is the bulk density.

The graph on the top in figure 24 illustrates the migration of uranium and breakthrough under saturated conditions in the < 2 mm fraction of sediment obtained from the 300 Area. Complete breakthrough occurred within the first five pore volumes. An effective retardation of 2.1 was calculated from moment analysis and afforded an apparent K_d of 0.29. Alternatively, complete breakthrough of uranium in a column containing a 5-wt% apatite barrier located in the center of the column did not occur until nearly 200 pore volumes had been displaced through the column (figure 24, bottom). Approximately a 100-fold increase in sorption was observed, $R_{ef} = 104.1$ and $K_{d-ap} = 22.97$.

In addition to conducting column tests to evaluate the effectiveness of an apatite barrier to sequester uranium, batch sequestration tests were conducted over a narrow pH range (6.0 to 8.0) to quantify the efficacy of uranium retention by apatite, under conditions comparable to the 300 Area.

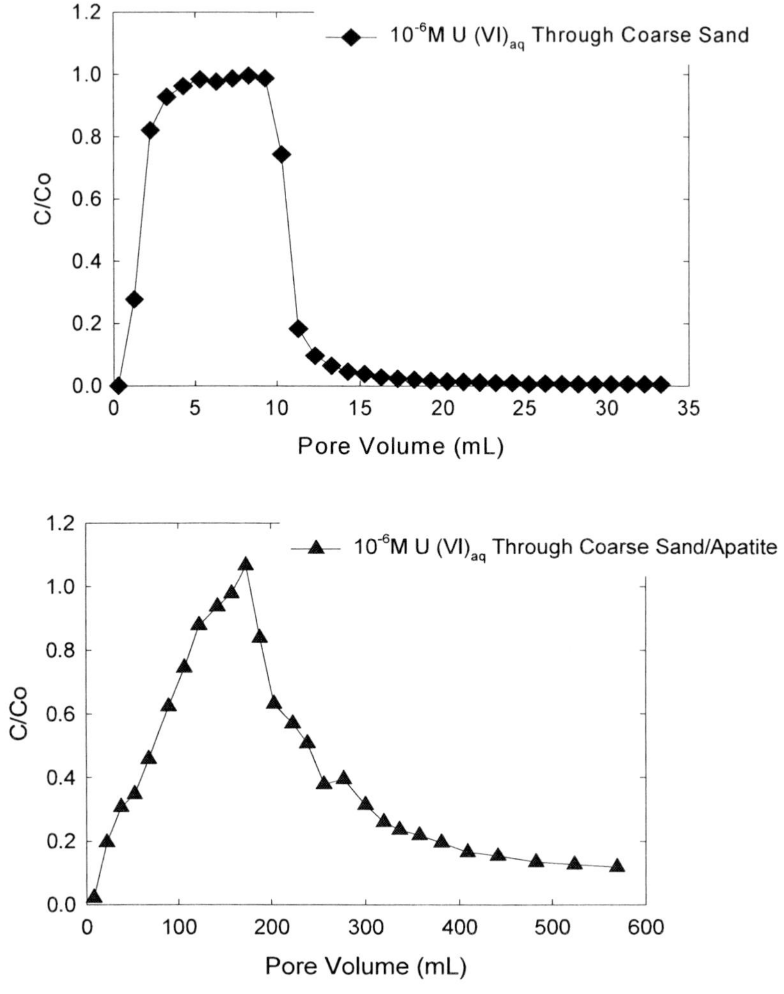

Figure 24. Experimental Migration of Uranium in Hydraulically Saturated Columns Through 300 Area Sediment (top) and Sediment Containing an Apatite Barrier (bottom).

The results of this investigation quantify the effect of pH on 1) the rate of uranium sorption on apatite, 2) the stability of uranium sorbed to apatite, and 3) measure the capacity of apatite for uranium sequestration. Moreover, the data illustrate the significance of surface properties and aqueous speciation on the sequestration of uranium by hydroxylapatite (Wellman et al. *in press*).

Chapter 9

APATITE AND AUTUNITE LONGEVITY

The Hanford subsurface does not contain a sufficient amount of naturally occurring phosphate to maintain phosphate mineral formations such as autunite and apatite, which will be in a state of thermodynamic disequilibrium. Consequently, it is necessary to understand the processes that will determine their longevity. The long-term stability of minerals is controlled by the solubility and the dissolution rate of the mineral. Under highly advective conditions, where transport is greater than the solubility, the stability of the mineral is controlled by dissolution kinetics. Alternatively, in low to moderately advective environments, solubility is greater than transport and the long-term stability of the mineral is based on the solubility of the phase. The former conditions are relevant to the 300 Area saturated zone. Therefore, it is necessary to evaluate apatite and autunite dissolution under conditions representative of the 300 Area aquifer. The dissolution rates of hydroxylapatite and autunite were quantified using the average flow rate within the 300 Area aquifer (Waichler and Yabusaki 2005).

QUANTIFICATION OF HYDROXYLAPATITE DISSOLUTION

Although phosphate minerals are important in natural processes such as biomineralization, control of soil nutrients, and industrial processes such as scale formation, phosphate mineral dissolution has not been investigated in detail. Until recently, the majority of the research has been conducted under conditions relevant to biology and focused on compositional analogues for skeletal apatite (Guidry and Mackenzie 2000). However, mineral composition and environmental conditions can control the mechanism and rate of dissolution. This is particularly noted for apatite minerals, such as fluorapatite [$Ca_5(PO_4)_3F$], hydroxylapatite

[$Ca_5(PO_4)_3OH$], and carbonate apatite [$Ca_5(PO_4, CO_3)_3(OH, F)$], because significant cation and/or anion substitution can occur during the formation and long-term weathering of these phases. The environment within the 300 Area subsurface ranges from pH 6 to 7.5, contains 0.80 ppm carbonate, 21.0 ppm chloride, and 0.60 ppm fluoride. As such, all major apatite isomorphs of apatite, hydroxylapatite, fluorapatite, and chlorapatite [$Ca_{10}(PO_4)_6Cl_2$], have the potential to form with varying degrees of cation and/or anion substitution.

Guidry and Mackenzie (2003) previously quantified the dissolution of fluorapatite and carbonate fluorapatite as a function of pH (2 to 8.5) and temperature (23° to 55°C). Results of this investigation illustrated that the dissolution of fluorapatite and carbonate fluorapatite were highly dependent on pH, and that the dissolution of fluorapatite (ranging from 10^{-9} mol m^{-2} s^{-1} at pH 5 to 10^{-10} mol m^{-2} s^{-1} at pH 8) was faster than carbonate fluorapatite (ranging from 10^{-11} mol m^{-2} s^{-1} at pH 5 to $10^{-11.5}$ mol m^{-2} s^{-1} at pH 7). Although there are no known investigations regarding the dissolution rate of chlorapatite, the heat of dissolution for chlorapatite lies between that for fluorapatite and hydroxylapatite (Stolyarova 1999) (figure 25). A conservative estimate of the longevity of the apatite barrier should be limited by the formation and stability of the least stable phase, hydroxylapatite.

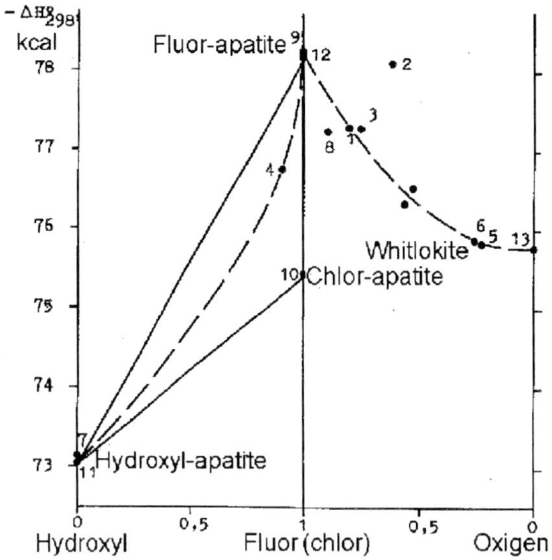

Figure 25. Heat of Dissolution for Fluor-, Chlor-, and Hydroxylapatite (Stolyarova 1999).

Results of numerous investigations of hydroxylapatite dissolution have provided valuable information regarding the mechanisms of hydroxylapatite corrosion as a function of relevant environmental variables, including temperature, pH, and solution species (Guidry and Mackenzie 2003; Dorozhkin 1997a, 1997b, 2002; Guidry and Mackenzie 2000; Kohler et al. 2005; Misra 1991; de Leeuw 2004; Valsami-Jones et al. 1998; Schaad et al. 1997, 1994; Christoffersen 1980; Christoffersen and Christoffersen 1982). Results of static dissolution studies identified the significance of surface processes during the dissolution of hydroxylapatite. Gramain et al. (1989, 1987) and Thomann et al. (1990) noted dissolution processes wherein limited mass transfer afforded the spontaneous formation at the solid interface of an adsorbed calcium-rich layer. Chistoffersen (1980) noted that under the pH range of $6.6 < pH < 7.2$ and $0.1 < C/C_s < 0.7$ where C is the stoichiometric concentration of hydroxylapatite, and C_S is the molar solubility of hydroxylapatite at the respective pH, the dissolution of hydroxylapatite is controlled by surface processes. Hydrogen ions catalyze the exchange of phosphate between the crystal surface and the solution (Christoffersen and Christoffersen 1982). Valsami-Jones et al. (1998) clarified the mechanism of dissolution and described two different surface groups: $\equiv Ca - OH_2^+$ and $\equiv P - O^-$, the latter of which is analogous to the $\equiv Si - O^-$ group found in silicates. Results of dissolution experiments conducted by Valsami-Jones et al. (1998) suggested deprotonation occurred at both sites, and within the pH range 5 to 7, $\equiv P - O^-$ was the significant site of deprotonation. In both instances, the charged species polarizes and weakens the phosphate bonds, $\equiv P\text{-}O^-$, ultimately leading to detachment of the phosphate molecule and subsequent degradation of the apatite structure. Results presented by Schaad et al. (1997) supported the significance of interfacial ionic exchange and calcium accumulation, which afford a pseudo-steady-state dissolution of hydroxylapatite. Therefore, the long-term performance of an apatite barrier will be determined by the rate of dissolution of the least stable potential apatite phase to precipitate during polyphosphate remediation—hydroxylapatite—as well as the rate of mass transfer. Depending on the rate of dissolution and precipitation, mass transfer may limit the degradation of an apatite barrier such that rapid reprecipitation of apatite will sustain the barrier beyond what would be possible if conditions were conducive to the forward rate of dissolution.

Figure 26 presents the release of calcium and phosphorus as a function of time over the pH range of 7 to 12, at a temperature of 40°C.

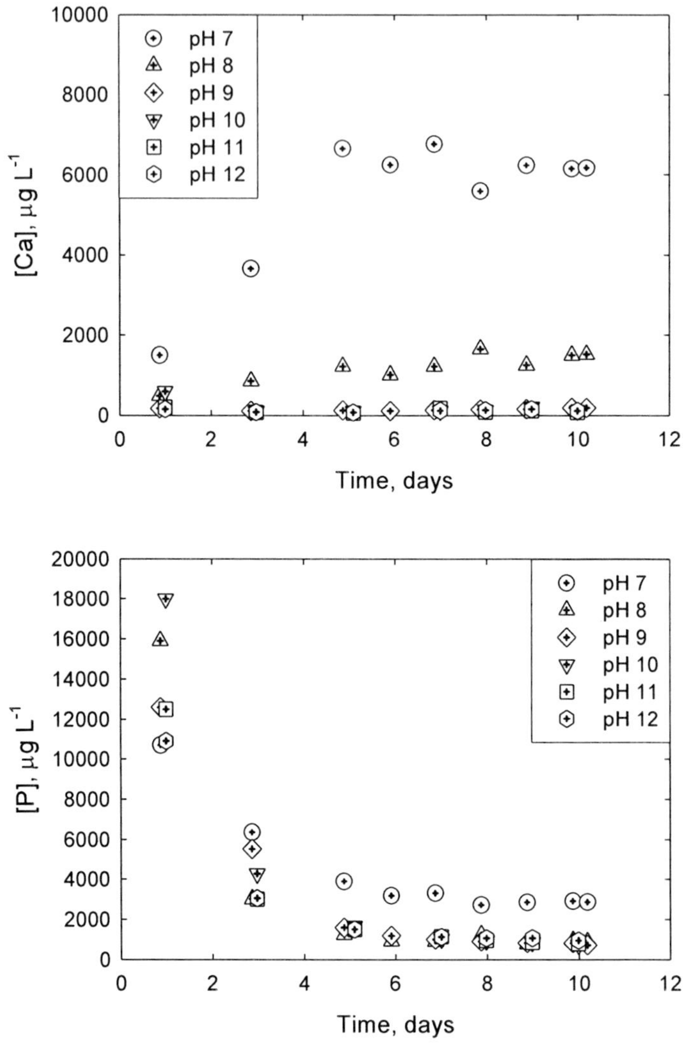

Figure 26. Aqueous Effluent Concentration of Calcium (top) and Phosphorus (bottom) as a Function of Reaction Time.

The graphs illustrate that steady-state conditions are met for calcium and phosphorus under all temperatures investigated. The results are representative for dissolution experiments conducted under the pH range of 7 to 12 at temperatures ranging from 23°C to 90°C. The concentration of calcium (figure 26, top) initially started out at low values and increased to steady-state values (concentrations invariant with time); whereas, the release of phosphorus (figure 26, bottom) initiated at high values and rapidly decreased to steady-state values.

The maximum dissolution rate for a given mineral is referred to as the forward rate of reaction or forward dissolution rate, which is typically attained by maximizing the ratio of flow rate (q) to surface area (S). This minimizes the concentrations of elements released into solution and decreases the chemical affinity within the system. This effect is expressed mathematically as the chemical affinity of the following reaction:

$$ChemicalAffinity, A, = RT \ \ln\left(\frac{K}{Q}\right) = -\Delta G_r \qquad (9)$$

where

ΔG_r = the free energy of reaction
R = the gas constant
T = the temperature
K = the equilibrium constant
Q = the ion activity product.

In other words, the chemical affinity is a measure of the departure from equilibrium. Therefore, as the ion activity product, Q, approaches the value of the equilibrium constant, K, the chemical affinity term goes to zero, and the dissolution rate slows as the difference in chemical potential between the solid phase and the solution decreases. By increasing the flow rate or decreasing the surface area within the system, the value of q/S can be increased, which increases the difference in chemical potential between the solid phase and the solution. This allows quantification of the maximum ratio of q/S necessary to attain the dissolution plateau, which is equated to the forward rate of dissolution (Nagy 1995). Accordingly, once the forward rate of reaction has been quantified, the effects of various environmental factors (e.g., effect of temperature, pH, and saturation state) on dissolution rate can be quantified independently by conducting experiments under the respective conditions to preclude dissolution rates from being influenced by solution saturation state.

Figure 27 illustrates the effect of varying the ratio of q/S. Dissolution rates based on the steady-state concentration of calcium (figure 27, top) and phosphorus (figure 27, bottom) are plotted for the conditions of 40°C, pH (23°C) = 9. At high q/S values (i.e., dilute conditions), the dissolution rate as indexed by calcium (figure 27, top) appears to be independent of solution saturation state and at the maximum (or forward) rate of reaction [3.5 x 10^{-12} mol m^{-2} s^{-1} at $\log_{10}$ (q/S) = -10.10] (Aagaard and Helgeson 1982). However, the dissolution rate indexed by phosphorus is, within experimental error, invariant as a function of q/S (figure 27, bottom), which could suggest that the solution saturation state was oversaturated with respect to secondary phases.

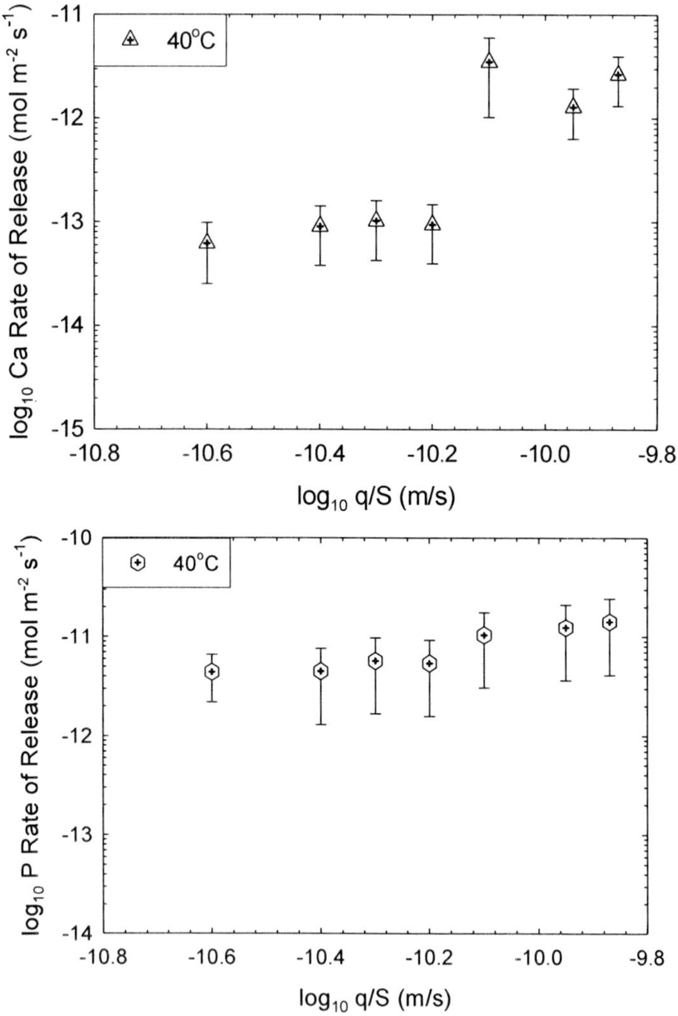

Figure 27. Effect of the Variation in q/S as Indexed by the Release of Calcium ($\log_{10}$) (top), and Phosphorus ($\log_{10}$) (bottom), from Hydroxylapatite.

The dissolution profiles when indexed based on the release of calcium (log_{10}) (figure 28, top) exhibited a negative pH dependence (η = -0.36) and no dependence on temperature. The apparent dissolution rate as indexed by the release of phosphorus (log_{10}) from hydroxylapatite displayed a similar negative dependence on pH over the pH range of 5 to 7. Under pH values greater than 7, within experimental error, there was no measurable dependence on pH or temperature (figure 28, bottom).

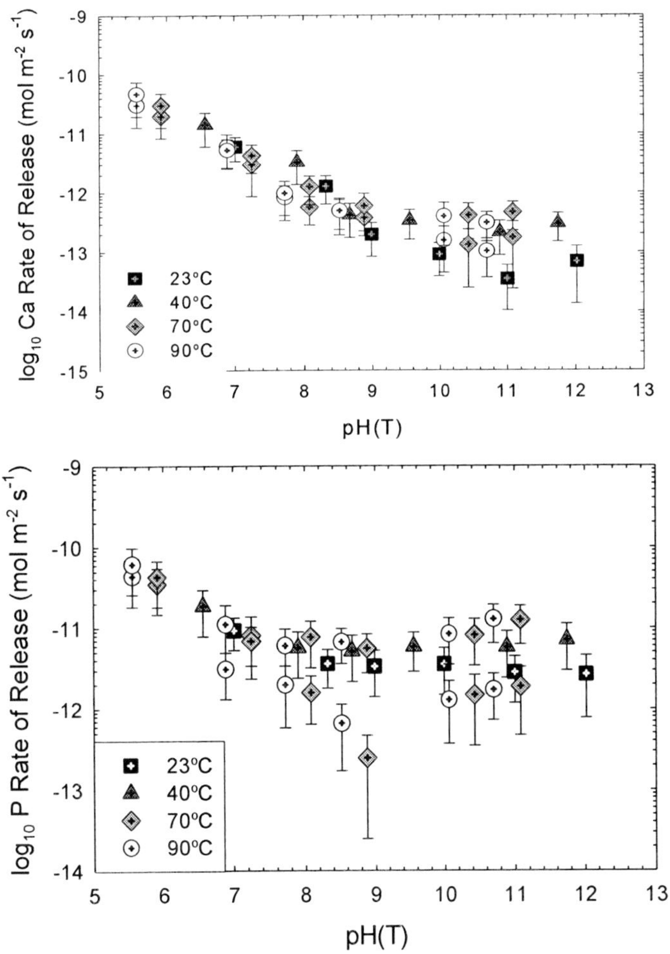

Figure 28. Rate of Hydroxylapatite Dissolution as Indexed by the Release of Calcium (log_{10}) (top) and Phosphorus (log_{10}) (bottom) as a Function of Temperature-Corrected pH

Figure 29 displays the average ratio of calcium to phosphorus as a function of pH as measured in the effluent solutions during dissolution. Within experimental error, the ratio of calcium to phosphorus is stoichiometric at pH 6 and 8. However, the ratio of Ca:P at pH 7 is 1.97. The presence of more calcium than phosphorus in solution suggests either 1) formation of a calcium-rich surface layer is inhibiting the release of phosphorus or 2) the aqueous concentration of phosphorus is being controlled by secondary phase precipitation.

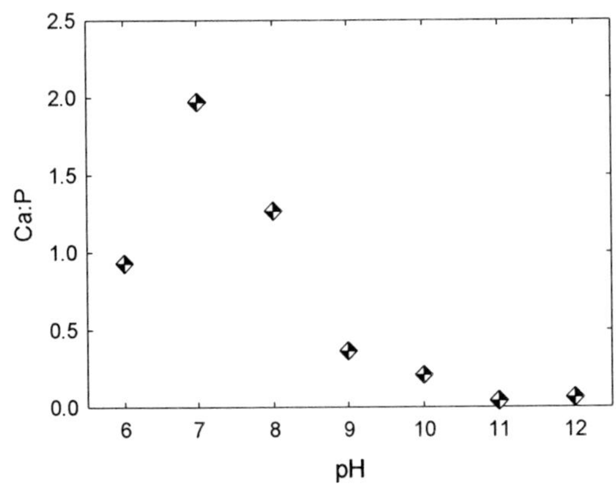

Figure 29. The Ca:P ratio as a Function of pH as Measured in Effluent Samples During Dissolution Testing.

Geochemist's Workbench®, version 3.2.2, (Bethke 1992) was used to predict the solid-phase stability based on the concentrations of calcium and phosphorus in the effluent solution. Figure 30 indicates that, given the concentration of calcium and phosphorus in solution, the system is undersaturated with respect to any calcium-phosphate solid phases. Similar analyses for conditions at 23°, 70°, and 90°C were conducted and yielded comparable results. This suggests quantification of the dissolution rate of hydroxylapatite is not confounded by the formation of calcium-phosphate solid phases. However, geochemical modeling of solution concentrations to predict solid-phase saturation implies the formation of secondary solid phases is via homogeneous precipitation from solution. This neglects the potential catalytic effects afforded by interactions with the dissolving solid phase, such as heterogeneous nucleation, topotaxy, and epitaxy, all of which have been shown to play a significant role in the formation of silicate minerals during dissolution (Nagy 1995; Putnis and Putnis 2007) and phosphate phase formation (Harlov et al. 2002; Yanagisawa et al. 1999; Renden-Angeles et al. 2000; Manecki et al. 2000).

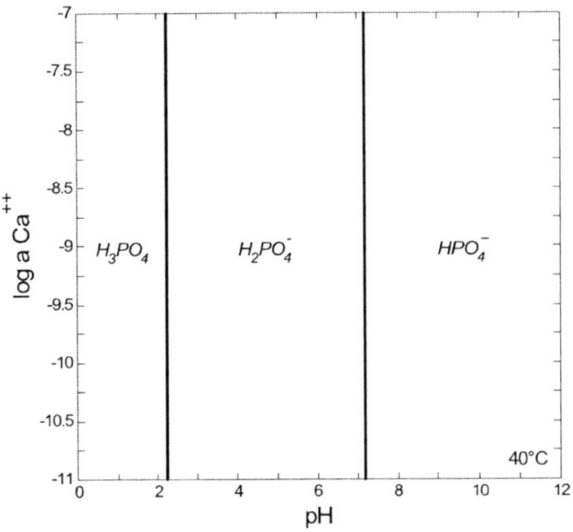

Figure 30. Predicted Saturation State Based on Effluent Solution Concentrations for the Dissolution of Hydroxyapatite in 0.05 M THAM at 40°C.

Although, hydroxylapatite is frequently the most thermodynamically stable solid phase predicted to control the activities of calcium and phosphate (Lindsay 1979; Lindsay and Moreno 1960), natural waters remain oversaturated with respect to hydroxylapatite due to the exceedingly slow rate of hydroxylapatite precipitation (Inskeep and Silvertooth 1988). Numerous investigations regarding the precipitation of hydroxylapatite have demonstrated the efficacy of seed crystal precipitation for the formation of hydroxylapatite (Brown 1981a; Ferguson et al. 1970; Inskeep and Silvertooth 1988; van Cappellen and Berner 1991; Nancollas et al. 1979; Nancollas and Tomazic 1974; Tomazic and Nancollas 1975; Moreno and Varughese 1981; Koutsoukos et al. 1980; Aoba and Moreno 1985; Boskey and Posner 1976; Kato et al. 1997; Amjad et al. 1981). The presence of seed crystals limits spontaneous nucleation of amorphous precursor phases (i.e., dicalcium phosphate dehydrate and octacalcium-phosphate) (Grossl and Inskeep 1992). As such, precipitation of hydroxylapatite can occur under conditions at low saturations with respect to hydroxylapatite (Koutsoukos et al. 1980; Moreno and Varughese 1981; Aoba and Moreno 1985; Boskey and Posner 1973; Kato et al. 1997).

Figure 31 illustrates the saturation state of calcium-phosphate minerals during the reaction of hydroxylapatite at 40°C. The input concentration of phosphorus and calcium were set to values measured in effluent solution samples, as discussed above. In contrast to results of geochemical modeling of solution chemistry, the presence of hydroxylapatite crystals significantly decreases the aqueous concentration of calcium and phosphorus necessary to promote precipitation of hydroxylapatite. Additionally, the system is saturated with respect to the thermodynamically less stable calcium-phosphate phase whitlockite [$Ca_9(Mg,Fe^{2+})(PO_4)_6(PO_3OH)$], which is frequently a precursor to the formation of hydroxylapatite.

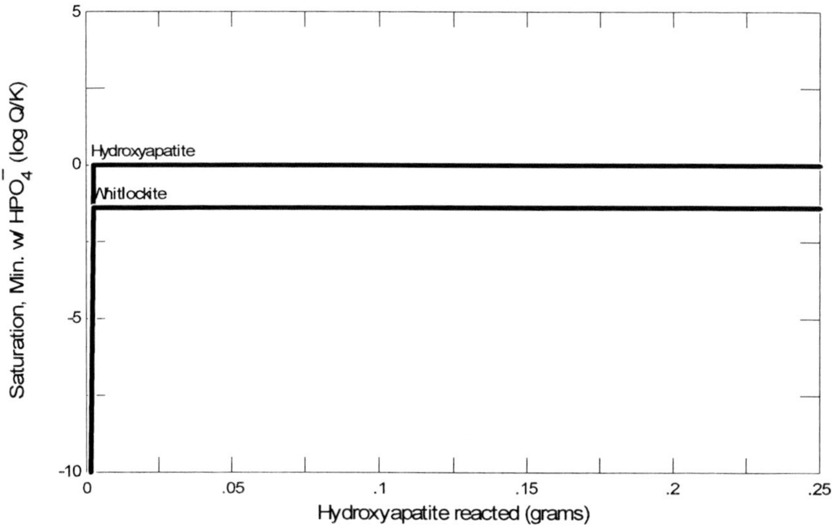

Figure 31. Predicted Phosphorus Mineral Saturation Indices Based on the Reaction of Hydroxylapatite at 40°C, pH 8.

Scanning electron microscopy conducted on hydroxylapatite following dissolution testing (T = 40°C, pH 8) clearly illustrates precipitation on the surface of the reacted starting phase (figure 32) . The fact that XRD of the reacted material does not reveal any reflections other than those attributable to hydroxylapatite suggests that during the dissolution of hydroxylapatite, the dissolving surface also acts as a nucleating surface for the precipitation of hydroxylapatite. Comparable results were observed during the analysis of reacted hydroxylapatite from all temperature and pH values investigated. Thus, given the average flow rate within the 300 Area aquifer, the performance of an apatite barrier may be sustained beyond that predicted based on solubility or dissolution kinetics due to reprecipitation. At pH values ≥ 9 the ratio of Ca:P is less than one (figure 29), suggesting that secondary precipitation of calcium-rich secondary phases may be controlling the aqueous concentration of calcium.

Figure 32. Scanning Electron Photomicrograph of Hydroxylapatite Reacted at 40°C, pH = 8, Illustrating the Re-Precipitation of Hydroxylapatite on the Surface of the Starting Material.

Figure 33 displays the saturation state with respect to calcium-bearing phases during the dissolution of hydroxylapatite at 40°C. In addition to the calcium-phosphate phases, saturation indices for lime, portlandite, and calcium hydroxide are exceeded under the test conditions.

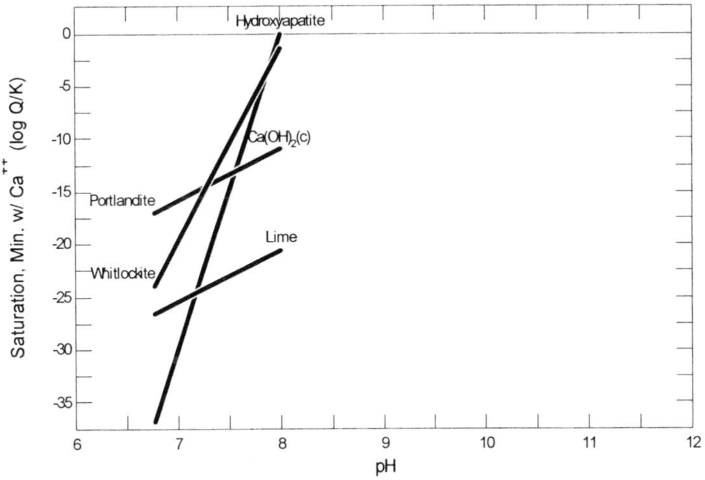

Figure 33. Predicted Calcium Mineral Saturation Indices Based on the Reaction of Hydroxylapatite at 40°C.

Rapid precipitation and dissolution of these phases, which are more soluble than hydroxylapatite, may influence the apparent dissolution rate such that a plot to evaluate the forward rate of dissolution appears to have attained the maximum rate of dissolution when, in fact, the system is being strongly influenced by rapid precipitation/dissolution of secondary solid phases (figure 27). No evidence of these phases was observed during SEM analyses. However, formation of secondary phases typically proceeds through initial formation of an amorphous phase. This precludes identification via XRD and, depending on the rate of reaction, may not be visible using SEM due to rapid redissolution or transformation to more stable crystalline phases. The only distinct crystalline phase, aside from the starting material, was hydroxylapatite in post-test material.

DYNAMIC WEATHERING OF HYDROXYLAPATITE

A column experiment was conducted to evaluate the stability of an in situ hydroxylapatite barrier. Figure 34 displays the release of calcium and phosphorus in Hanford groundwater at 23°C.

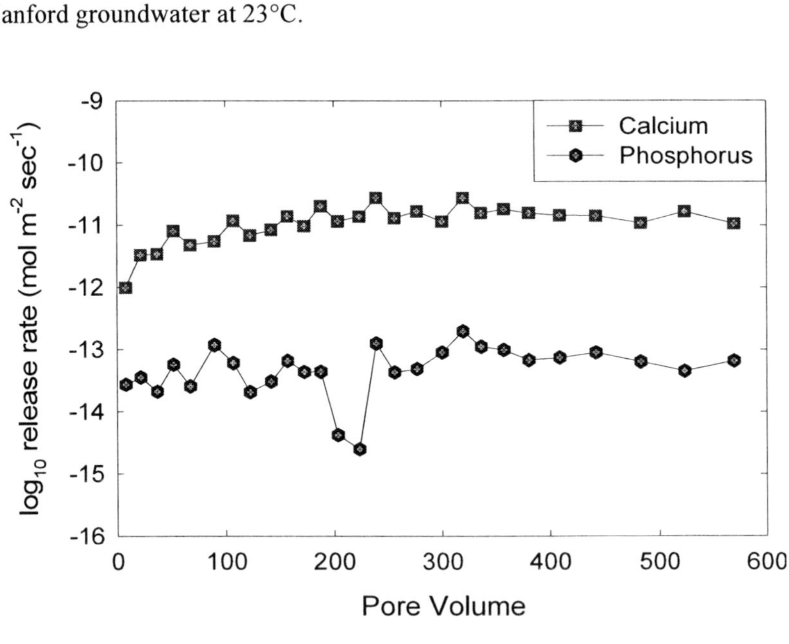

Figure 34. Experimental Release of Calcium and Phosphorus from 300 Area Sediment Containing a 5-wt% Hydroxylapatite Barrier.

The apparent release rate as indexed by the release of calcium and phosphorus (log_{10}) from hydroxylapatite indicates rapid attainment of steady state. Unlike results obtained by SPFT testing, wherein the calculated release rates for calcium and phosphorus were within experimental error, the release of calcium (1.42×10^{-11} mol m^{-2} sec^{-1}) from the sedimentary column experiment was approximately 1000 times faster than that measured for phosphorus (7.14×10^{-14} mol m^{-2} sec^{-1}). Moreover, the release of calcium measured here was within experimental error of that quantified by SPFT testing; however, the release of phosphorus was approximately two orders of magnitude lower than SPFT results.

It has been noted previously that the evolution of the hydroxylapatite surface is responsible for attainment of pseudo-steady-state rather than true steady-state, which indicates that the decreased rates of dissolution as indexed by phosphorus were due to the formation of a semi-permeable layer of interfacial calcium accumulation (Gramin et al. 1989; Gramin et al. 1987; Thomann et al. 1990). Moreover, thermodynamic equilibrium of hydroxylapatite may be reached under non-stoichiometric conditions (Schaad et al. 1997). Results of dynamic weathering tests for hydroxylapatite indicate it is likely that dissolution/reprecipitation reactions will sustain an in situ apatite barrier. This will serve as a long-term sorbent for uranium and a source of phosphate for the formation of uranium-phosphate phases under the geochemical and hydraulic conditions present within the 300 Area.

DISSOLUTION KINETICS OF META-AUTUNITE MINERALS, $(Na, Ca)_{2-1}[(UO_2)(PO_4)]_2 \cdot 3H_2O$

Uranyl-phosphate phases are advanced secondary uranium minerals formed during the oxidized weathering of primary UO_2 deposits (Garrels and Christ 1965). The general paragenetic sequence of secondary mineral formation has been well-documented (Finch and Murakami 1999); however, uranyl- silicates and -phosphates are typically the solubility-limiting minerals that persist in locales geographically removed from the primary deposit (Murakami et al. 1997). The presence of phosphate in groundwater, even in minor concentrations (10^{-8} M), promotes the formation of autunite group minerals $X_{3-n}^{(n)+}$ $[(UO_2)(PO_4)]_2 \cdot xH_2O$; thereby, limiting the mobility of the uranyl cation (UO_2^{2+}) in subsurface environments.

Previous experimental results have established the low solubility of many uranyl-phosphate minerals (Moskvin et al. 1967; Vesely et al. 1965; Chukhlantsev and Stepanov 1956; Scheyer and Baes 1954; Karpov 1961). However, knowledge of the stability of the uranyl-phosphate phases is restricted to a narrow range of experimental conditions involving low pH media with high concentrations of phosphoric acid (Scheyer and Baes 1954; Vesely et al. 1965; Karpov 1961). Further, all known studies related to autunite stability and solubility are based on synthetic, rather than natural, phases (Giammar 2001; Sowder et al. 2000; Vesely et al. 1965; Pekarek and Vesely 1965; Scheyer and Baes 1954; Karpov 1961), and interpretations of dissolution studies have been confounded by impurities present within the starting material (Giammar 2001).

In situ stabilization of uranium as autunite via the injection of sodium polyphosphate and calcium chloride (for subsequent formation of apatite) may result in the formation of Na-autunite and/or Ca-autunite. As such, evaluating the efficacy of polyphosphate technology requires quantifying the long-term stability of uranium immobilized within autunite, be it Na- or Ca-autunite. Figure 35 (top) illustrates the release rate of uranium from Na-autunite, across the pH range 7 to 10 and the temperature range of 5° to 70°C. Release rates of uranium increase by ~100 fold over the pH interval of 7 to 10. Under the temperature range of 5° to 40°C the increase in rate as a function of pH is constant, and is quantified as the power law coefficient, $\eta = 0.91 \pm 0.08$. The constant value of the slope over the temperature interval indicates that the power law coefficient, η, is independent of temperature. However, the constant release of uranium as a function of pH deviates at the higher pH values of 9 and 10 at 70°C. This suggests the formation of a secondary phase(s), which would result in slower release of uranium at conditions of high pH and temperature.

Geochemical modeling suggests that experiments with Na-autunite are undersaturated with respect to all possible secondary phases over the pH range of 7 to 10 from 5° to 40°C. However, at 70°C the system becomes saturated with respect to schoepite, β - $UO_2(OH)_2$, α - $UO_3{\cdot}9H_2O$, at pH 9 and 10, and, additionally, clarkeite (sodium uranyl oxy-hydroxide) at pH 10. These results support the validity of autunite dissolution rates obtained over the temperature interval 5° to 40°C. However, the results at 70°C indicate that the concentration of uranium is not due solely to dissolution of Na-autunite; secondary phases control net uranium concentrations.

Figure 35 (bottom) depicts the release rate of phosphorus across the pH range 7 to 10 over the temperature range of 5° to 70°C. The release of phosphorus is less dependent on pH ($\eta = 0.51 \pm 0.17$), than observed for uranium ($\eta = 0.91 \pm 0.08$); yet, the rate of phosphorus release is faster by approximately 30 times. Release of phosphorus follows the same pH- and temperature-dependent patterns exhibited by uranium. Over the temperature range of 5° to 40°C the increase in rate as a function of pH is constant; however, at 70°C, the slope of the line decreases. Yet, geochemical modeling did not suggest saturation with respect to any phosphorus-bearing phases. This suggests that dissolution of the autunite structure is dependent on initial removal of uranium from the uranium-phosphate sheet.

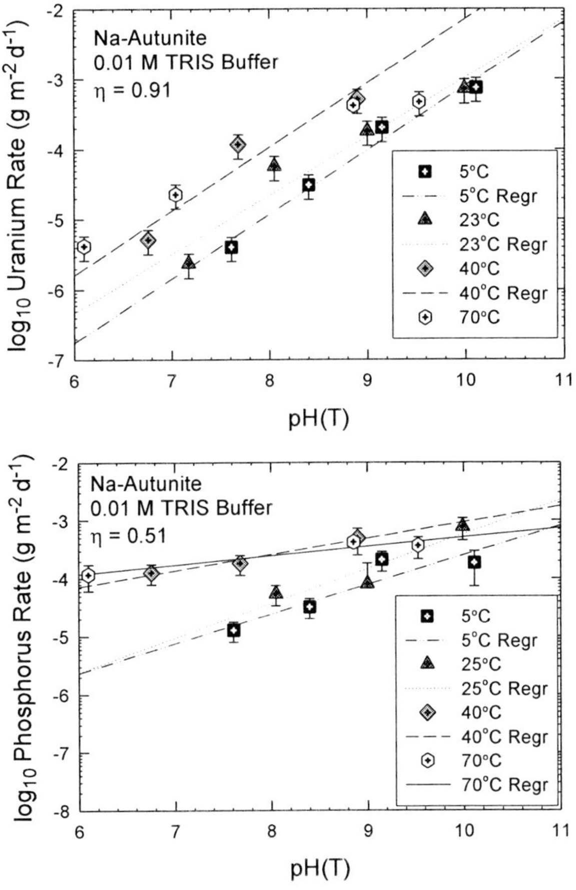

Figure 35. The $\log_{10}$ Uranium Release Rate as a Function of Temperature-Corrected pH for Na-Autunite in 0.01 M TRIS solution (top), and the $\log_{10}$ Phosphorus Release Rate as a Function of Temperature-Corrected pH for Na-Autunite in 0.01 M TRIS Solution (bottom).

The pH dependence for uranium release, η = 0.88 ± 0.04, measured across the pH range 7 to 10 (5° to 40°C) for GHR was identical within error to the release of uranium from Na-autunite, η = 0.91 ± 0.08 (figure 35 (top) and figure 36 (top)). An inflection in pH dependence was again observed at 70°C, suggesting saturation limits with respect to secondary uranyl minerals may also be controlling the release of uranium in GHR tests.

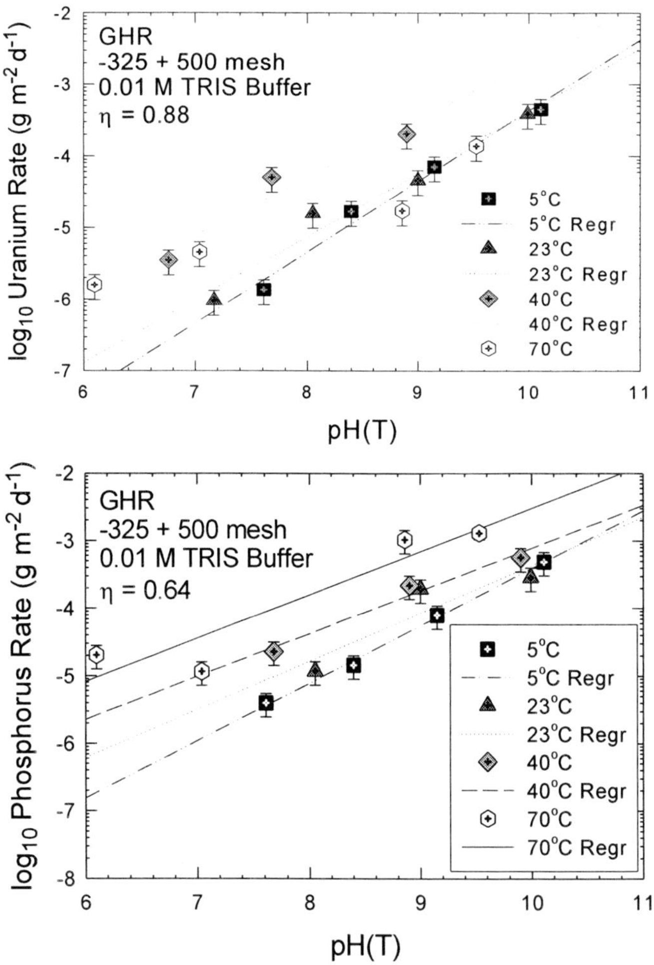

Figure 36. The $\log_{10}$ Phosphorus Release Rate as a Function of Temperature-Corrected pH for Na-Autunite in 0.01 M TRIS Solution (top), and the $\log_{10}$ Phosphorus Release Rate as a Function of Temperature-Corrected pH GHR in 0.01 M TRIS Solution (bottom).

Table 8. Mineral Saturation Indices Based on Effluent Solution Compositions from Dissolution of GHR in 0.01 M TRIS Buffer

Temp., °C	pH	Mineral Saturation Indices					
		Schoepite	β - $UO_2(OH)_2$	α - $UO_3 \cdot 9H_2O$	$CaUO_4$	Hydroxylapatite	Whitlockite
5	7-9	Undersaturated with respect to all potential secondary minerals					
	10	–	–	–	2.608	4.151	-0.238
23	7-9	Undersaturated with respect to all potential secondary minerals					
	10	–	–	–	3.696	4.648	0.096
40	7-8	Undersaturated with respect to all potential secondary minerals					
	9	–	–	–	1.094	–	–
	10	–	–	–	4.517	6.228	1.079
70	7-8	Undersaturated with respect to all potential secondary minerals					
	9	–	–	–	2.756	3.134	-0.187
	10	-0.122	-0.067	-0.229	4.984	8.245	2.640

Table 8 summarizes the geochemical modeling results based on steady-state effluent concentrations for GHR. The presence of calcium has a noticeable effect on the chemical affinity of the system. At pH = 10, the system becomes saturated with respect to $CaUO_4$ across the temperature range investigated. At the two highest temperatures investigated, 40° and 70°C, the system becomes saturated with respect to $CaUO_4$ at pH = 9. The pattern of element release rates, coupled with geochemical modeling results, explain the convergence of apparent uranium release rates to a common value, 3.57×10^{-4} g · m-^{2} d^{-1} at pH = 10, across the investigated temperature range. Results at 70°C indicate the system also becomes saturated with respect to schoepite, β - $UO_2(OH)_2$, and α - $UO_3{\cdot}0.9H_2O$ at pH 10, which is similar to that predicted for Na-autunite. Thus, uranium dissolution data from 5° to 40°C are representative of the dissolution of GHR. However, uranium release rates at 70°C are subject to solubility limits for secondary minerals, and cannot be attributed solely to dissolution of GHR.

Phosphorus release rates from GHR are shown in figure 36 (bottom). In accordance with phosphorus release from Na-autunite, the pH-dependent release of phosphorus was less than that quantified for uranium, 0.64 ± 0.04 versus 0.88 ± 0.03, respectively. However, release of phosphorus from GHR exhibits no deviation in phosphorus release with increasing temperature and pH, which is contrary to the pattern of phosphorus release displayed by Na-autunite (c.f. pH 9 and 10 at 70°C). This suggests that secondary mineral solubility may be influencing the apparent phosphorus concentrations. Geochemical modeling results suggest that the system exceeds the saturation index for hydroxylapatite at pH 10 under all temperatures investigated. Thus, the apparent release of phosphorus at all temperatures and high pH is not solely attributable to the dissolution of GHR under these conditions.

Interlayer Cation Release Rates

Contrary to uranium and phosphorus release rates, which were shown to increase as a function of pH and also with increasing temperature (figure 35), release of sodium is shown here to be independent of pH and temperature. Release of sodium is ~9,800 times faster than uranium at the lower pH values of 7 and 8 (figure 37, top). As pH increases, the difference in the release rate of sodium relative to uranium is significantly less; sodium release is only ~7 times greater than uranium at pH 10 (figure 37, top). Moreover, release rates for calcium from GHR display similar characteristics to sodium release from Na-autunite, in that both are independent of temperature and pH (figure 37). Comparing release of calcium from GHR (bottom) to sodium release from Na-autunite (top) illustrates that the rates are identical within experimental error (figure 37).

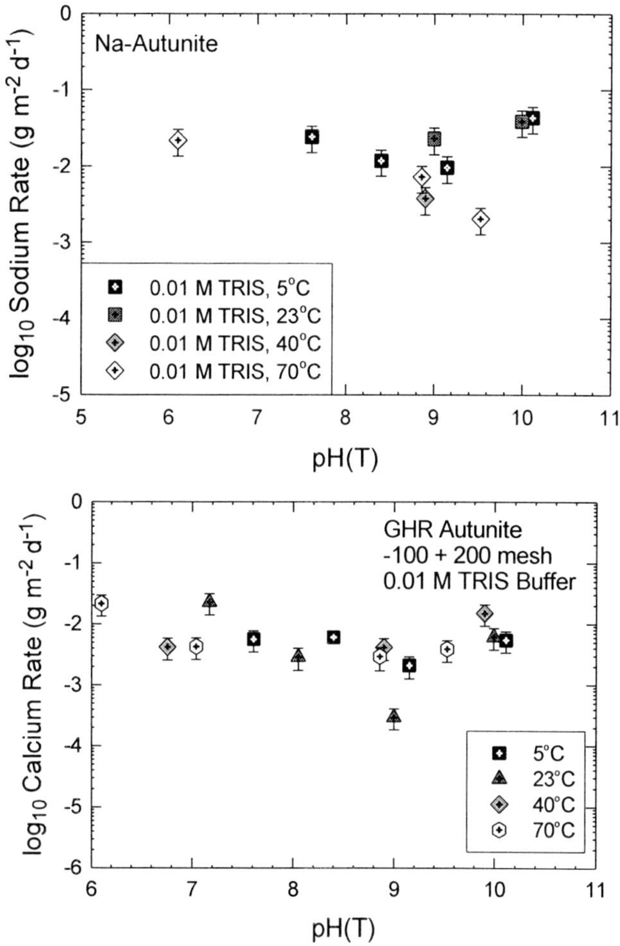

Figure 37. The $\log_{10}$ Sodium Release Rate as a Function of Temperature-Corrected pH for Na-Autunite in 0.01 M TRIS solution (top), and the $\log_{10}$ Calcium Release Rate as a Function of Temperature-Corrected pH for GHR in 0.01 M TRIS Solution (bottom).

Release of interlayer cations (i.e., Na^+ or Ca^{2+}) from minerals is generally subject to two separate reactions: matrix dissolution and alkali-hydrogen exchange. Based on the saturation state of the system, one or both of these mechanisms may contribute to release of interlayer cations from the structure. For example, when the system is near saturation, the activities of dissolved species near and/or in contact with the solid phase increase, thereby resulting in a decrease in the matrix dissolution rate. Concurrently, the chemical potential difference between autunite and solution will be the driving force for cation diffusion. Concentrations of both Na^+ and Ca^{2+} appear to be constant across the range of pH values, but this is misleading. Dissolution of the autunite matrix will also contribute to the concentration of dissolved cations in solution; therefore, two distinct mechanisms, ion exchange and matrix dissolution, account for Na^+ or Ca^{2+} release.

STRUCTURAL DISSOLUTION

SPFT experiments suggested that the dissolution of autunite occurs via attack and removal of the uranium polyhedra. To provide a more thorough understanding of the autunite dissolution mechanism, select SPFT experiments were conducted at 40°C in D_2O-based solutions. The test solution was a 0.1 M deuterated ammonium hydroxide (ND_4OD) solution. The pD of the solution was adjusted using deuterated hydrochloric acid (DCl). The pH and pD scales are related through the following equation:

$$pD = pH + 0.4 \tag{10}$$

Steady-state release rates in H_2O and D_2O-based solutions are listed in table 9. Uranium release rates from both autunite minerals are approximately an order of magnitude slower in D_2O than quantified in H_2O. In contrast, phosphorus rates are statistically within error between H_2O and D_2O-based solutions.

The decrease in uranium release rates with no effect on phosphorus release rates supports the hypothesis that the dissolution of autunite minerals is controlled by a surface-mediated reaction with the uranium polyhedra. The mean bond enthalpy of H_2O is 463.5 kJ mol^{-1} compared to 470.9 kJ mol^{-1} for D_2O reflecting the greater strength of the D_2O bond compared to H_2O. The observed decrease in the release rate of uranium in D_2O reflects a surface complex that requires the breakage of an O-H (O-D) bond. Thus, the slower reaction rates in D_2O are contributed to by a rate-limiting step in the hydrolysis of uranium within the autunite sheet.

Table 9. Release Rates from Autunite Samples in H_2O and D_2O at pH(D) = 8, T = 40°C

Autunite	Solution Composition	U Release Rate ($g \cdot m^{-2} d^{-1}$)	P Release Rate ($g \cdot m^{-2} d^{-1}$)	Na/Ca Release Rate ($g \cdot m^{-2} d^{-1}$)
Na-Autunite [$Na_2(UO_2)_2(PO_4)_2 \cdot 3H_2O$]	H_2O	2.3×10^{-6} ($\pm 2.1 \times 10^{-5}$)	1.41×10^{-5} ($\pm 7.3 \times 10^{-6}$)	9.4×10^{-3} ($\pm 4.4 \times 10^{-3}$)
	D_2O	8.3×10^{-8} ($\pm 2.2 \times 10^{-8}$)	3.8×10^{-5} ($\pm 7.3 \times 10^{-6}$)	3.2×10^{-3} ($\pm 5.6 \times 10^{-4}$)
GHRI [$Ca(UO_2)_2(PO_4)_2 \cdot 3H_2O$]	H_2O	2.7×10^{-7} ($\pm 4.5 \times 10^{-8}$)	2.8×10^{-5} ($\pm 1.8 \times 10^{-5}$)	2.8×10^{-3} ($\pm 1.1 \times 10^{-3}$)
	D_2O	7.2×10^{-8} ($\pm 2.7 \times 10^{-8}$)	5.3×10^{-6} ($\pm 2.0 \times 10^{-6}$)	1.8×10^{-3} ($\pm 7.0 \times 10^{-4}$)

Release rates of interlayer cationic elements show no dependence on the identity of the cation or variation in release rate based on the solution media (i.e., H_2O versus D_2O). This is significant because if the rates of release were governed solely by ion exchange with H^+ or H_3O^+ an isotopic effect would be detectable and proportional to the square root of the ratio of the masses of the isotopic atoms, $\sqrt{\frac{H}{D}} = 0.71$. Thus, a 71% difference in the release would be detected in the results.

Experiments conducted in D_2O-based buffer solutions exhibited a decrease in uranium release rates by approximately an order of magnitude, relative to rates calculated in H_2O-based buffer solutions, while phosphorus rates were consistent between buffer systems. This supports a proposed mechanism in which autunite dissolution proceeds via attack at the uranium polyhedral units by •OH, rather than at the phosphate tetrahedra.

The autunite structure is characterized by perfect (*001*) basal cleavage with relatively weak forces holding successive sheets together, thereby increasing the probability that dissolution of the autunite mineral could occur through structural attack by water molecules along cleavage planes. Separation of the autunite sheets during dissolution would readily release the interlayer cations into solution. Na-autunite material used in dissolution experiments did not exhibit any cleavage planes prior to dissolution. However, SEM analyses of reacted Na-autunite revealed the formation of cleavage planes during the dissolution process (figure 38). This supports the proposed hypothesis that dissolution occurs through attack of the crystal from the edges and along the cleavage planes. Additionally, this affords a significant contribution to the release of interlayer cations. Thus, interlayer cation release behavior is a combination of structural dissolution and ion exchange, but imparts no effect on the overall stability of autunite.

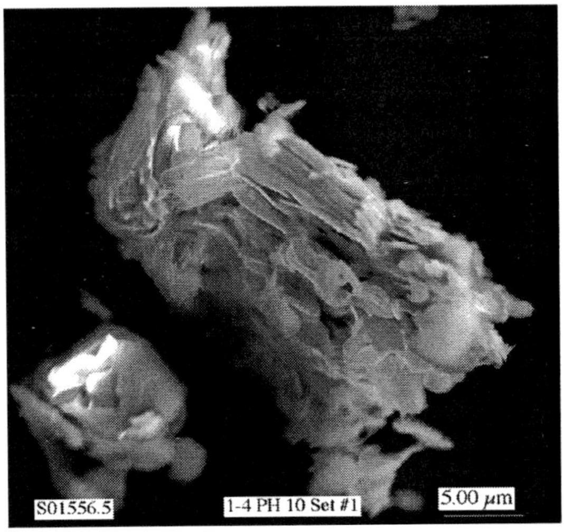

Figure 38. A SEM Photomicrograph of Reacted Na-Autunite Illustrating Basal Cleavage of the Autunite Plates from Attack During Dissolution.

Chapter 10

PILOT-SCALE FIELD STUDIES

WELL ARRAY AND HYDROGEOLOGY

The well array for the test consisted of one injection well and eight monitoring wells (figure 12). The uppermost unconfined aquifer beneath the polyphosphate treatability test site consists of unconsolidated gravel-dominated sediments of the Hanford Formation unit 1. The Hanford-Ringold contact boundary ranged in depth between approximately 13.1 and 15.2 m (43 and 50 ft) below ground surface (bgs) (table 10). Typical depths to water range from 10.6 m (35 ft) bgs during low river stage conditions to 9.1 m (30 ft) bgs during spring/summer high-water conditions, resulting in a Hanford aquifer thickness of 4.5 to 6 m (15 to 20 ft). In all of the wells, from ~10.9 m (36 ft) bgs to the Ringold Formation contact, the Hanford Formation consists predominantly of coarse sandy gravel to gravel (figure 39). A more open framework, i.e., clast-supported structure; composed of predominantly gravel to slightly sandy gravel (figure 39) is reported in Wells 399-1-24, 399-1-26, 399-1-30, and 399-1-31 in the lower Hanford Formation from approximately 10.9 m (36 ft) bgs down to the Hanford-Ringold contact; where present, the matrix sand is composed of medium to coarse sand.

The Ringold Formation unit 5 that lies beneath the Hanford Formation is composed of mostly gravelly silty sand to sand. With the exception of two shallow well pairs and the earlier Well 399-1-23 (which was drilled down to the lower mud unit) , the wells were drilled a few feet into the Ringold Formation to a total depth of ~15.5 m (51 ft) bgs (table 10); therefore, only the upper few feet of Ringold Formation sediments were encountered. At all locations, the Hanford-Ringold contact was distinguished by a distinct color change, decrease in gravel size and content, and a significant increase in fine sand.

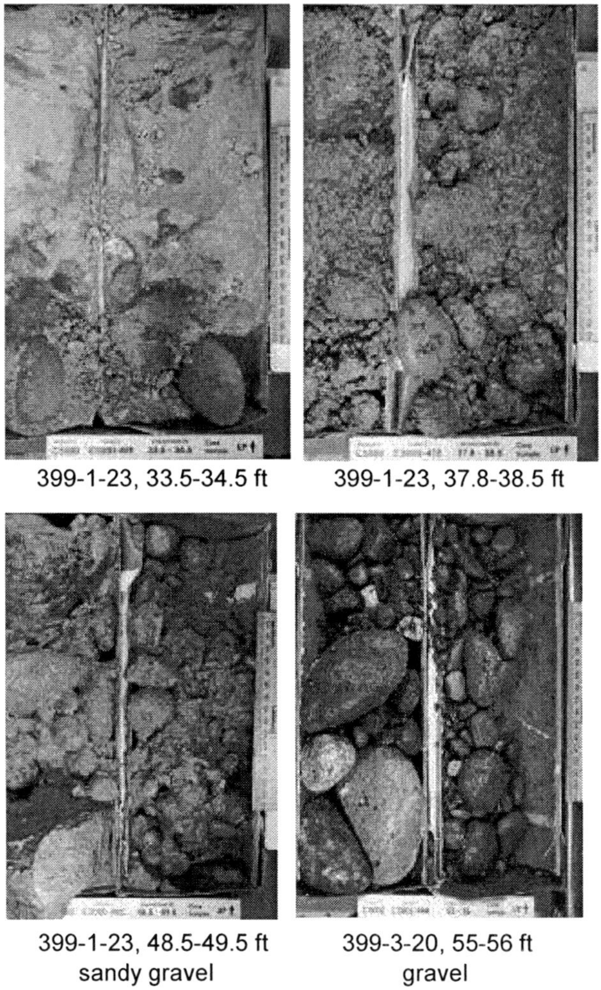

Figure 39. Photos of Silty-Sandy gravels (top) and Open Framework Gravels (bottom) Representative of the Hanford Formation.

The eastern-most well, Well 399-1-32, encountered the Hanford-Ringold contact was ~1.2 to 1.5 m (4 to 5 ft) higher in elevation than the other new wells (table 10). Approximately 2.3 m (7.5 ft) of fairly uniform, Ringold Formation sand with only minor amounts of gravel was recovered from the core barrel below the contact. The contact appears fairly well defined in this well and strongly suggests that the Ringold Formation sand encountered in the other wells reflect the same unit.

WATER TABLE AND GROUNDWATER FLOW DIRECTIONS

The water table at the 300 Area is very dynamic due to fluctuations in the Columbia River stage (figure 40) and leads to the very high permeability of the Hanford Formation sediments (table 11) comprising the uppermost part of the unconfined aquifer. Large daily, weekly, and seasonal fluctuations in the Columbia River stage are caused by the operation of hydroelectric dams on the river and seasonal trends (i.e., spring freshet). The dynamics of river stage fluctuations and the water table elevation cause a mixing zone of river and groundwater within the aquifer. During relatively high river stage periods river water enters the aquifer.

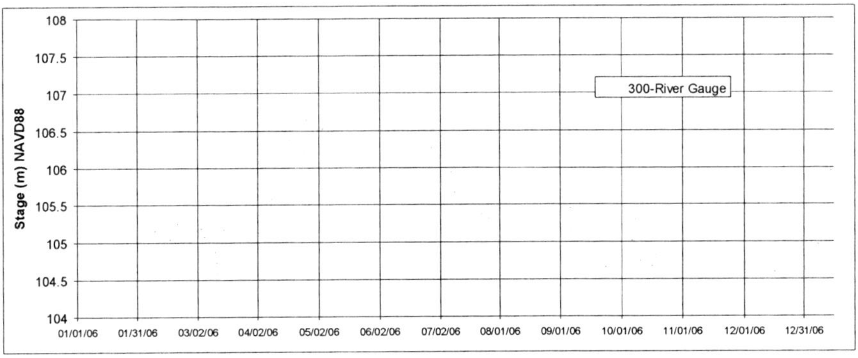

Figure 40. Columbia River Stage from the 300-Area Gauging Station for 2006.

Measurements of specific conductance and temperature in wells in the 300 Area, where the groundwater and river water have a large contrast in values, show that river water can encroach more than 190 m inland in the aquifer during a high river stage period. During relatively low river stage periods, groundwater discharges to the river, as indicated by specific conductance measurements (and other analytes) in aquifer tubes installed below the river bed and in springs along the shoreline (Patton et al. 2003).

Table 10. Polyphosphate Treatability Well Identification and Borehole Information

Well Name	Well ID	Surface Elevation (ft)	Total Drill Depth (ft bgs)	DTW (ft bgs)	H-R Contact (ft bgs)	Saturated Hanford fm. (ft)	Elevation H-R (ft)	Screen Interval/Length (ft bgs [and ft])
399-1-23[a]	C5000	378.84	116	34.5	51	16.5	327.84	25-50 (25 ft)
399-1-24	C5351	379.35	42	33.8	NDE	NDE	NDE	32-37 (5 ft)
399-1-25	C5352	379.27	50	34.3	48.2	13.9	331.07	42-47 (5 ft)
399-1-26	C5353	378.82	50.5	33.8	48.5	14.7	330.32	29-49 (20 ft)
399-1-27	C5354	379.59	50	34.9	48	13.1	331.60	42-47 (5 ft)
399-1-28	C5355	379.64	40.5	34	NDE	NDE	NDE	32-37 (5 ft)
399-1-29	C5356	379.60	51	35	49	14	330.60	29-49 (20 ft)
399-1-30	C5357	379.38	50.5	33.6	49.75	14.75	329.63	29-49 (20 ft)
399-1-31	C5358	379.03	51	33.7	48.5	14.8	330.53	29-49 (20 ft)
399-1-32	C5359	378.21	50.5	32.5	43	10.5	335.21	29-44 (15 ft)

(a) Pre-existing 6-inch-diameter injection well, all other wells are new and 4 inches in diameter. All screens completed with 20 slot wire-wrap stainless steel and 10- to 20-mesh silica sand pack.

bgs = below ground surface

DTW = down to watertable

H-R = Hanford-Ringold contact boundary

NDE = not deep enough.

Table 11. Physical Property Data for Bulk Sediment Samples from Well 399-1-23 (Williams et al. 2007)

Well ID	Sample	Elevation Mid-pt (m)	Unit	Bulk Density (g/cm^3)	Total Porosity[a]	% Grav	% Sand	% Silt	% Clay
399-1-23	C5000-39D	107.83	H	1.95	0.293	71.78	21.15	4.16	2.92
399-1-23	C5000-40C	105.69	H	2.34	0.152	76.18	19.43	3.02	1.37
399-1-23	C5000-40E	105.08	H	2.31	0.165	70.59	22.12	5.34	1.95
399-1-23	C5000-41C	104.47	H	2.34	0.153	76.45	19.73	2.55	1.26
399-1-23	C5000-45C	98.99	R	2.26	0.182	82.77	13.18	3.03	1.02

(a) Particle density was not measured so an average particle density = 2.76 g/cm^3 (see Williams et al. 2006, Table 3) was used to calculate porosities.

The water table for a low river period (December) and high river stage period (June) are shown in figure 41. The water table, along with the uranium concentrations, is distinctly different during these periods as shown in the figures. Additionally, the water table is relatively flat in the 300 Area (i.e., very small hydraulic gradients) due to the extremely high permeability of the Hanford Formation comprising the uppermost portion of the unconfined aquifer.

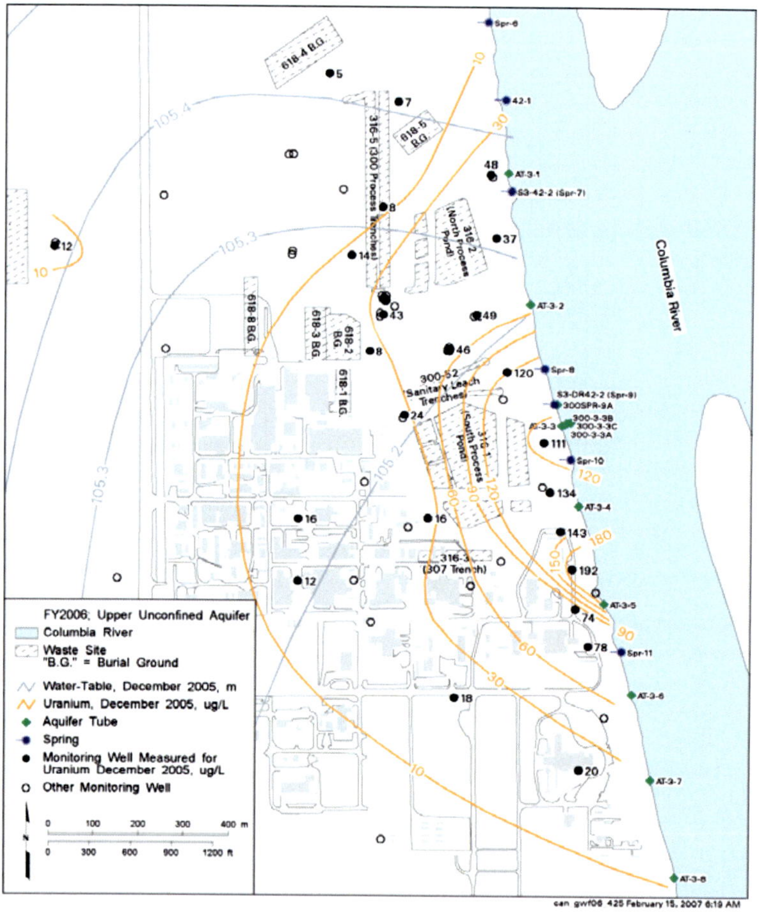

Figure 41. Continued on next page.

Figure 41. Water Table and Uranium Concentrations in Upper Part of Unconfined Aquifer Beneath 300 Area, December 2005 (top) and June 2006 (bottom) (Hartman et al. 2007).

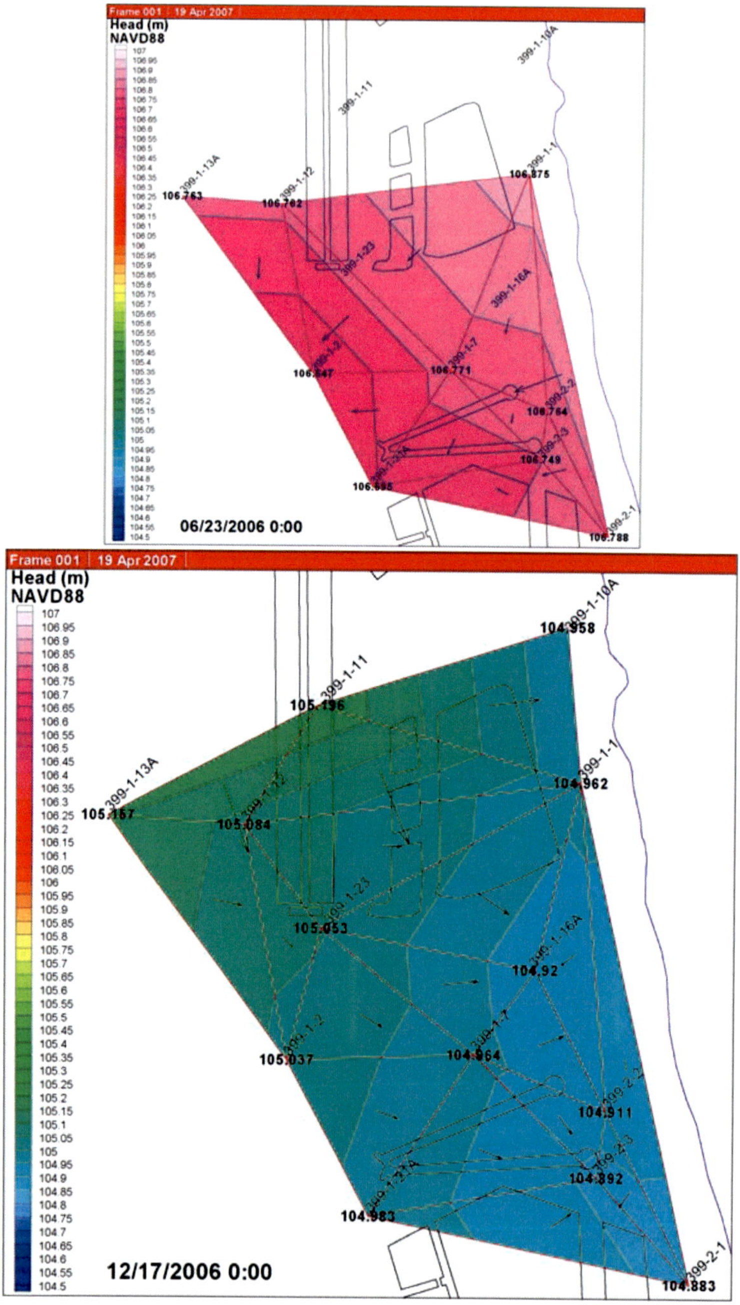

Figure 42. 300 Area Water-Level Monitoring Network – High River Stage (top), Low River Stage (bottom) (Vermuel et al. 2006).

Contoured hydraulic head data and calculated hydraulic gradients for two selected time periods, high and low river stage for 2006, are shown in figure 42. The higher resolution water level monitoring network, which includes the polyphosphate injection well was not operational during a high river stage period, so the coarser data set was used to guide downgradient directions during this period. As shown by figure 42, groundwater flow directions are inland during the June high river period and toward the river during the December low river stage period. Monthly rose diagrams showing the groundwater flow direction from this network using measurements every two hours are shown in figure 43.

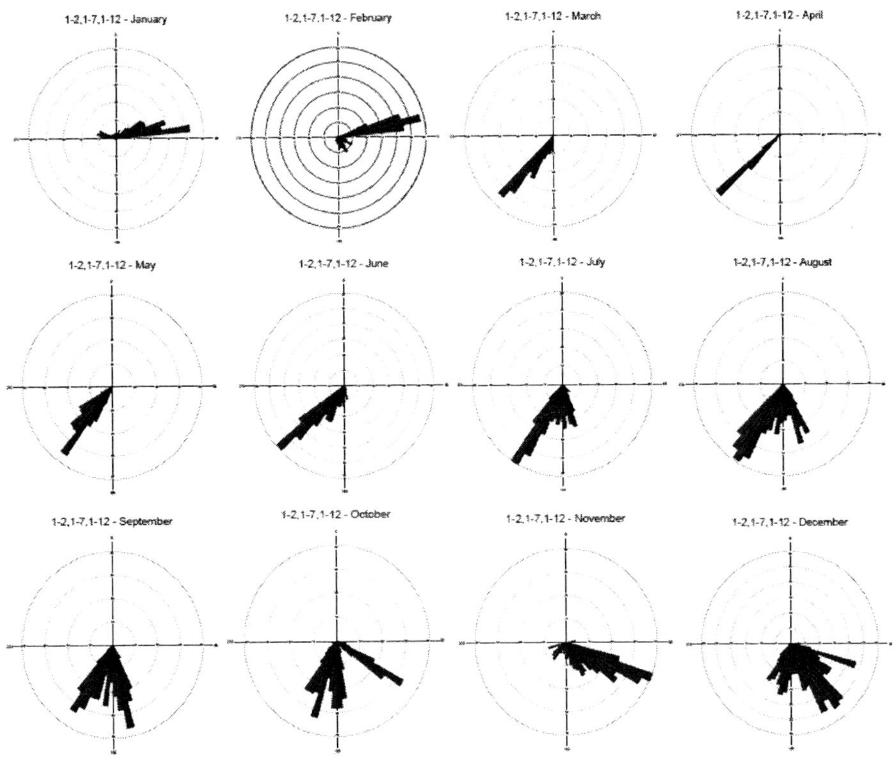

Figure 43. Hydraulic Gradient Directions for 2006 Calculated from 300-Area Automated Water-Level Network (two-hour data intervals used) for Well Cluster 399-1-2, 399-1-7, and 399-1-12. Azimuth shows direction towards flow (March 16, 2007 data) (Vermuel et al. 2006).

TRACER TEST RESULTS AND DISCUSSION

The tracer injection test results provide information on aquifer heterogeneities, effective porosity, expected arrival times and required solution volume for the polyphosphate injection. Bromide BTCs were constructed for all of the wells monitored during the test. The results are discussed in two groups: wells within the targeted injection volume (8.8 m [29 ft] radial extent) and downgradient wells.

Targeted Injection Volume Monitoring Wells

Within the targeted injection volume, 50% bromide concentration arrival times (t_{50}) ranged from 16 to 428 minutes (table 12). These results indicate a general correlation between tracer arrival time and radial distance from the injection well, with a few notable outliers. Wells 399-1-26 (northwest of the injection well) and 399-1-31 (northeast of the injection well) are both about 6 m (20 ft) from the injection well and had similar arrival times of 111 and 90 minutes, respectively (figure 44).

Table 12. Bromide Tracer Injection Arrival Times and Porosity Results for Targeted Injection Volume Monitoring Wells

Well Name	Well Screen Zone	Radial Distance (ft)	50% Tracer Arrival (min)	Average Velocity (ft/day)	Estimated Effective Porosity
399-1-23	Full	0	–	–	–
399-1-24	Upper	14.9	124	168	0.33
399-1-25	Lower	14.4	39	519	0.11
399-1-26	Full	19.9	111	260	0.16
399-1-27	Lower	24.5	428	81	NC
399-1-28	Upper	24.9	216	162	0.21
399-1-29	Full	29.6	165	254	0.11
399-1-30	Full	14.8	16	1300	–
399-1-31	Full	19.6	90	316	NC

NC = not calculated due to uncharacteristic response. Average = 0.18.

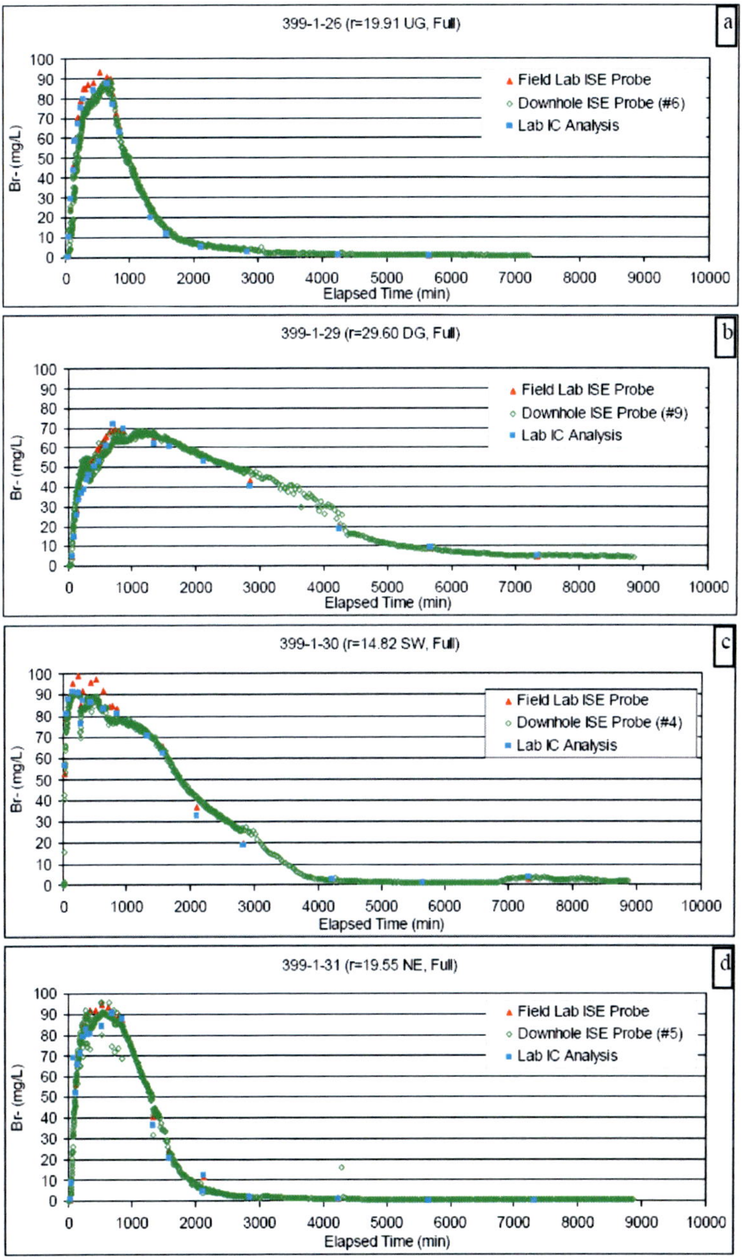

Figure 44. Breakthrough Curves (BTCs) Showing Bromide Concentrations Through Time for Fully Screened Wells Within the Targeted Injection Volume: a) 399-1-26, b) 399-1-29, c) 399-1-30, and d) 399-1-31.

Well 399-1-29, located on the perimeter of the targeted injection volume (radial distance of 8.8 m [29 ft]), reached concentrations of ~70 mg/L or ~80% of full concentration, indicating that this location was near the outer extent of the injection pore volume (PV) in this radial direction. Well 399-1-30 is an outlier among the other fully screened wells within the targeted injection volume, showing a much quicker arrival time than the other wells (t_{50} = 16 minutes). The observed early arrival at this location most likely is associated with formational heterogeneities resulting a preferential flow path between the injection well and this monitoring well location.

Assessment of the vertical variability in bromide tracer arrival was possible by comparing the results of adjacent wells screened in upper and lower zones of the aquifer (figure 45). Two upper/lower screen well pairs were installed on the downgradient side of the targeted injection volume (figure 12). The 399-1-24/399-1-25 well pair exhibited a similar peak concentration for both depth intervals but the 50% tracer arrival in 399-1-25 occurred in one-third the time, indicating preferential flow within the lower portion of the aquifer between these two locations.

The other upper/lower screen well pair, 399-1-27 and 399-1-28, showed very different arrival responses for the two intervals. The tracer arrival and peak concentration for 399-1-28, screened in the upper aquifer zone, is similar to other wells within the targeted injection volume. However, bromide arrival response in Well 399-1-27, which is screened in the lower zone, showed an unexpectedly slow arrival and low overall concentration at this monitoring location. The BTC shows that the peak bromide concentration in this well remained below 50% of the injection stream concentration over the duration of the injection (figure 45). The relatively low well yields provided by this lower zone monitoring well, suggest that the lower zone of the aquifer at this location is less transmissive than the upper zone at this location or the lower zone at the other well pair location.

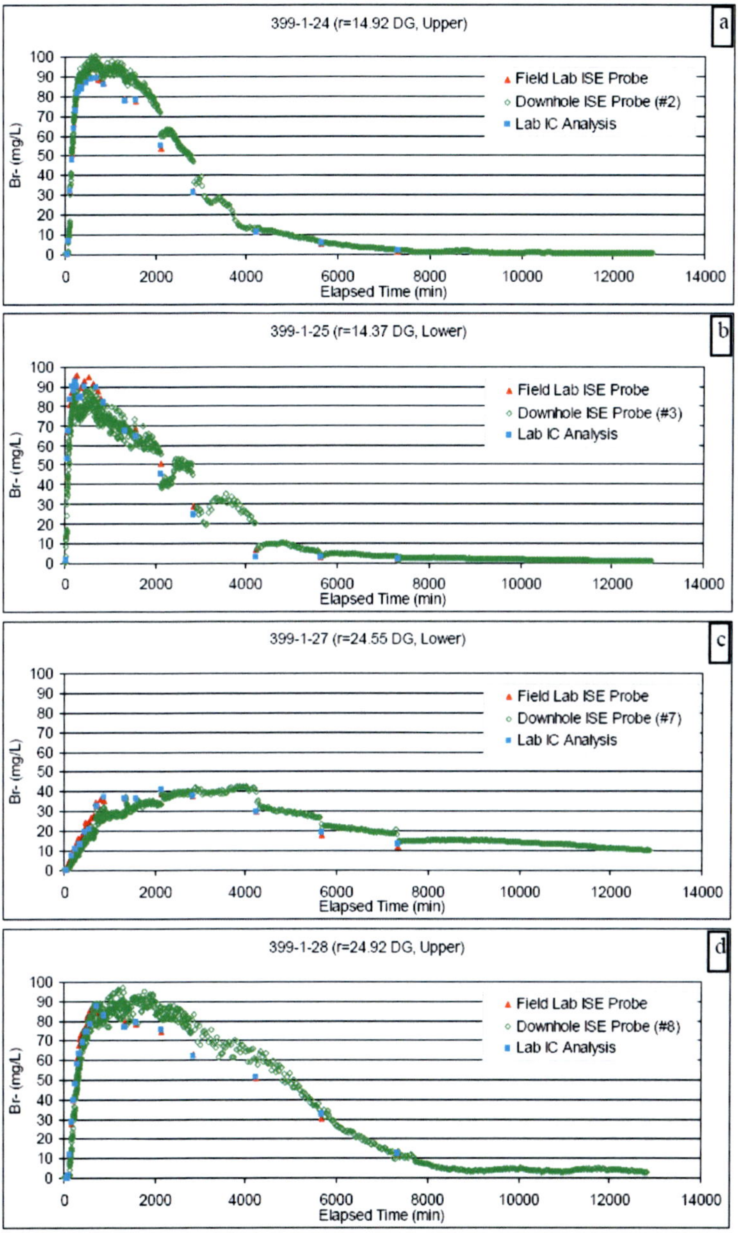

Figure 45. Breakthrough Curves (BTCs) Showing Bromide Concentrations Through Time for Wells Within the Targeted Injection Volume that are Screened in Only the Upper or Lower Zones of the Aquifer: a) 399-1-24, b) 399-1-25, c) 399-1-27, and d) 399-1-28.

The tracer arrival times were used to estimate the effective porosity of the aquifer according to the following equation:

$$n = \frac{t_{50} \cdot Q_{tot}}{\pi \cdot r^2 \cdot L \cdot 7.48} \tag{11}$$

where n = effective porosity
t_{50} = 50% Br^- concentration arrival time (minutes)
Q_{tot} = total injection rate (200 gpm)
R = radial distance from the injection well (feet)
L = aquifer thickness (15 feet).

Effective porosities were calculated for each of the eight monitoring wells in the targeted injection volume, except for the two outlier wells (table 12). Values range from 11 to 33% for the different wells, with an average effective porosity of 18%.

Downgradient Monitoring Wells

Several monitoring wells are located downgradient from the injection well beyond the radial extent of the targeted injection volume (figure 46). These include Wells 399-1-32 and 399-1-17, located 104 and 617 feet from the injection well, respectively. By combining the results from the bromide tracer drift with water level measurements, and the resulting hydraulic gradient calculations, it is possible to estimate hydraulic conductivity (*K*) according to Darcy's Law:

$$v = \frac{K}{n} \cdot \frac{dh}{dx} \tag{12}$$

$$K = \frac{v \cdot n}{\frac{dh}{dx}}$$

where K = hydraulic conductivity (ft/day)

v = groundwater velocity based on the tracer arrival time (ft/day) *n* = average effective porosity from the tracer arrival times (18% from above)

dh/dx = time-weighted average hydraulic gradient during tracer transport (ft/ft)

Figure 46. Polyphosphate Treatability Test Site Showing Well Layout and Downgradient Monitoring Wells.

For the main tracer plume arrival at Well 399-1-32, the groundwater velocity was estimated at ~15.24 m (50 ft) per day during tracer transport, based on a radial distance of 31.69 m (104 ft) and a tracer transport duration of ~3,000 minutes (table 13). The tracer drift duration was defined as the time period between the end of the test when the tracer plume was centered over the injection well (t = 714 minutes) and the arrival time of the center of mass at 399-1-32 (t = ~3,700 minutes). The time-weighted average gradient during tracer transport between the injection well and Well 399-1-32, as determined from water-level measurements, was ~6.5 x 10^{-4} ft/ft. The estimated hydraulic conductivity using these parameters is about 4,300 m (14,000 ft) per day.

Table 13. Summary of Parameters Used to Estimate Hydraulic Conductivity Between Injection Well 399-1-23 and Downgradient Well 399-1-32 for Main Body of Tracer

$Time_{start}$[1] (elapsed minutes)	$Time_{end}$[2] (elapsed Minutes)	Transport Duration[3] (minutes)	Groundwater Velocity (ft/day)	Time-weighted Average Hydraulic Gradient (ft/ft)	Hydraulic Conductiv ity (ft/day)
714	~3700	~3000	~50	~6.5E-4	~14,000

1. Time when tracer plume was centered on Injection Well 399-1-23. This was the end of the test.
2. Time when tracer plume arrives at 399-1-32.
3. Time duration of tracer transport between Injection Well 399-1-23 and Well 399-1-32; defined as $Time_{end} - Time_{start}$.

Note: An effective porosity value of 18% was used in calculating the hydraulic conductivity estimate.

PILOT-SCALE POLYPHOSPHATE INJECTION

The concentrations of uranium in samples obtained from the well array before and after the polyphosphate treatment are shown in figure 47. Prior to remediation with polyphosphate the aqueous concentration of uranium ranged from ~50 to 110 µg/L. Data obtained a week after the treatment indicated significant reduction in uranium concentrations in most wells within the targeted and extended treatment zones. The average post-treatment uranium concentration in wells within the treatment and extended treatment zone was 28 µg/L after one week and 19 µg/L after two weeks as compared to the average pretreatment concentration of 79 μg/L. Uranium concentrations in most of the wells in both targeted and extended treatment zone were below the MCL of 30 μg/L. The noted outliers are Wells 399-1-27 and 399-1-28.

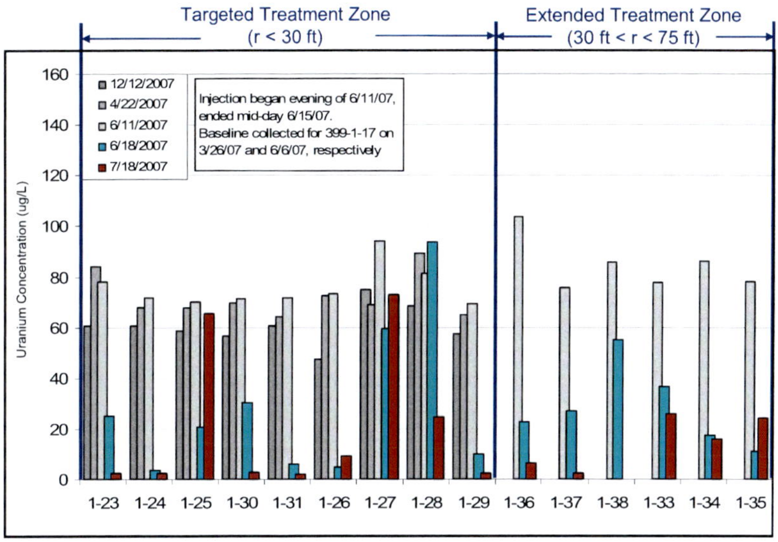

Figure 47. Pre- and Post-Treatment Concentrations of Uranium in Water Samples from the Well Array in the Polyphosphate Treatment Area.

The concentration of uranium arrival in Well 399-1-27 was only slightly reduced from that measured prior to treatment. Results of the preliminary tracer study, as discussed above, showed slow arrival of bromide and the overall concentration was low.

This suggested the lower zone of the aquifer at this location is less transmissive than the upper zone at this location or the lower zone at the other well pair location. These results suggest delivery of the polyphosphate amendment to the region near Well 399-1-27 was impeded. Well 399-1-28 also exhibited no reduction in uranium concentration one week subsequent to treatment with polyphosphate technology. However, five weeks after treatment with polyphosphate the concentration of aqueous uranium at Well 399-1-28 was below the MCL of 30 μg/L. This suggests heterogeneities delayed the arrival of the polyphosphate amendment at Well 399-1-28.

One week after treatment, water samples from wells screened in the lower zone (Well 399-1-25 and Well 399-1-27) showed reductions in uranium concentrations; however, at the end of five weeks, the concentrations had returned to the pretreatment levels. During borehole drilling the highest concentration of uranium was observed in the profile located just below the Hanford-Ringold contact, indicating that vertical gradients generated during liquid waste disposal activities at the 316-5 process trenches likely drove contamination into the uppermost portion of the Ringold Formation. Although the relatively low permeability of the Ringold Formation and cessation of liquid discharge to the trenches results in limited potential for additional transport of uranium within the Ringold Formation, these contaminated sediments will continue to act as a deep uranium source at this location. This conceptual model is supported by hydraulic conductivity values that varied greatly above and below the contact as defined by depth-discrete slug tests conducted during borehole drilling. Values of hydraulic conductivity in the saturated Hanford Formation from previous aquifer tests in the 300 Area are greater that 1000 m (3280 ft) per day andvalues in the Ringold Formation are in the range of tens of meters per day and less. These data suggest that most of the dissolved uranium moves through the permeable interval above the Hanford-Ringold contact. Additionally, the bromide BTC at Well 399-1-27 indicated that the peak bromide concentration in this well remained below 50% of the injection stream concentration over the duration of the injection (figure 45). The relatively low well yield likely limited the amount of polyphosphate amendment that permeated this zone providing limited initial response with rebound to background concentrations upon depletion of the minimal treatment capacity. Additional data collection and analysis is on-going to assess the post treatment uranium concentrations observed in these lower zones.

Excess orthophosphate was included within the amendment formulation to ensure precipitation kinetics were sufficient to overcome the high hydraulic conductivity and pore water velocities present within the open-framework of the 300 Area subsurface. Pressure response measurements were made in the injection well, Well 399-1-26, to assess any reduction in hydraulic conductivity engendered by the polyphosphate treatment. The data showed about 3- to 4-fold decreases in hydraulic conductivity indicating precipitation and formation of a reactive barrier in the target treatment zone (figure 48). These data were independently confirmed by tracer drift studies. It is evident from the results presented here that even under extreme pore water velocities and in high-conductivity media the polyphosphate technology can be engineered to provide lateral treatment of a contaminant plume and/or formation of a reactive barrier.

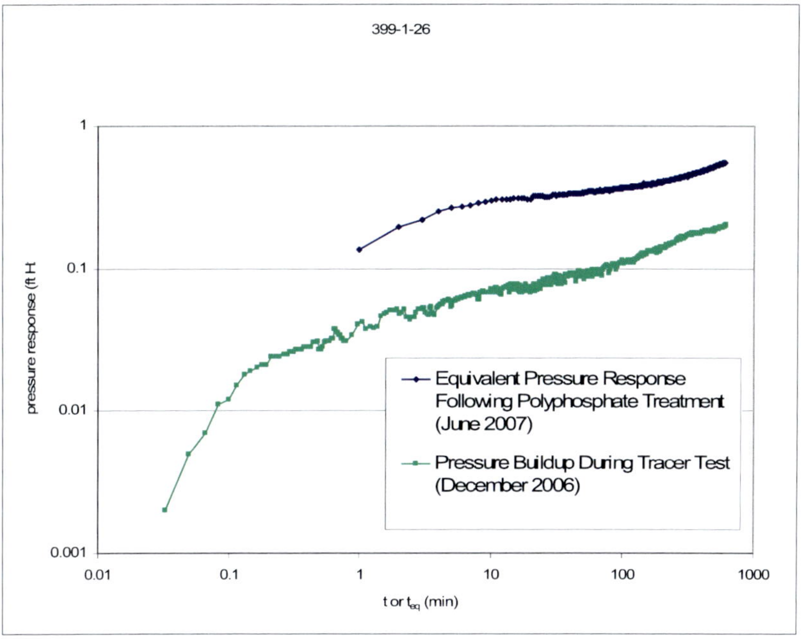

Figure 48. Pre- and Post-Treatment Pressure Response Data from Well 399-1-26.

Preliminary field data indicate that polyphosphate injection technology has the potential to effectively capture and sequester uranium from contaminated groundwater plumes resulting in the reduction of uranium concentrations well below the EPA MCL of 30 μg/L. Additional post-treatment analyses, including continued groundwater monitoring and core sampling, is on-going to assess the long-term effectiveness of the polyphosphate injection technology.

Chapter 11

SUMMARY AND CONCLUSIONS

Polyphosphate technology has been developed for controlled application and precipitation within deep subsurface contaminant plumes. Standard groundwater wells allow injection of polyphosphate technology deep below the ground surface providing treatment of contaminants too deep for conventional solid-phase apatite emplacement, or rapidly precipitating, soluble applications of short-chain inorganic or organic phosphate sources. The degradation kinetics of long-chain polyphosphate molecules, such as sodium tripolyphosphate, affords a time-released source of inorganic phosphate. Polyphosphate technology affords direct remediation of soluble uranium through the formation of an insoluble uranyl-phosphate mineral, autunite $X_{1-2}[(UO_2)(PO_4)]_{2-1}\bullet\ nH_2O$, where X is any monovalent or divalent cation. Because autunite sequesters uranium in the oxidized form, U(VI), rather than forcing reduction to U(IV), the possibility of re-oxidation and subsequent re-mobilization of uranium is limited. Release of uranium from autunite may only occur through dissolution of the mineral structure. In addition to autunite, excess phosphorous can result in apatite (calcium phosphate) mineral formation, providing a long-term source of treatment capacity. A field-demonstration has been conducted within a uranium contaminated aquifer on DOE's Hanford Site in southeastern Washington State. Results presented here underscore the necessity to understand the hydrogeology of the treatment site and the geochemistry controlling uranium within the subsurface environment. A thorough understanding of site-specific geochemical and hydrodynamic conditions enables optimization of polyphosphate technology for controlled remediation without discharging excess phosphorus into the environment, which could detrimentally impact the environment or downgradient drinking water supplies. Preliminary field data indicate that the polyphosphate injection technology has the potential to effectively capture and sequester uranium from contaminated groundwater plumes resulting in the reduction of uranium concentrations well below the EPA MCL of 30 μg/L

References

Aagaard, P, and H Helgeson. 1982. "Thermodynamic and Kinetic Constraints on Reaction Rates among Minerals and Aqueous Solutions, I. Theoretical Considerations." *American Journal of Science* 282:237-285.

Allison, JD, DS Brown, and KJ Novo-Gradac. 1991. *Minteqa2/Prodefa2, A Geochemical Assessment Model for Environmental Systems: Version 3 User's Manual*, EPA/600/3-91/021, Environmental Research Laboratory, Office of Research and Development, U.S. Environmental Protection Agency (EPA), Athens, Georgia.

Allison, JD, DS Brown, and KJ Novo-Gradac. 1998. *Minteqa2/Prodefa2, A Geochemical Assessment Model for Environmental Systems: User Manual Supplement for Version 4.0.*, Environmental Research Laboratory, Office of Research and Development, U.S. Environmental Protection Agency (EPA), Athens, Georgia.

Alwan, AK, and PA Williams. 1980. "The Aqueous Chemistry of Uranium Minerals. Part 2. Minerals of the Liebigite Group." *Mineralogical Magazine* 43:665-667.

Amjad, Z, PG Koutsoukos, and GH Nancollas. 1981. "Crystallization of Fluoroapatite. A Constant Composition Study." *Journal of Colloid and Interface Science* 82(2):394-400.

Aoba, T, and EC Moreno. 1985. "Adsorption of Phosphoserine onto Hydroxylapatite and Its Inhibitory Activity on Crystal Growth." *Journal of Colloid and Interface Science* 106:110-121.

Arey, JS, JC Seaman, and PM Bertsch. 1999. "Immobilization of Uranium in Contaminated Sediments by Hydroxylapatite Addition." *Environmental Science and Technology* 33:337-342.

American Society for Testing and Materials (ASTM). 2002. *Standard Test Method for Particle-Size Analysis of Soils*, D422-63, ASTM International, West Conshohocken, Pennsylvania.

Bethke, CM. 1992. *The Geochemist's Workbench*. Urbana-Champaign, IL: University of Illinois.

Boskey, AL, and AS Posner. 1973. "Conversion of Amorphous Calcium Phosphate to Microcrystalline Hydroxylapatite. A pH-Dependent, Solution-Mediated, Solid-Solid Conversion." *Journal of Physics and Chemistry* 77:2313-2317.

Boskey, AL, and AS Posner. 1976. "Formation of Hydroxylapatite at Low Supersaturation." *Journal of Physics and Chemistry* 80:40-45.

Brown, JL. 1980. "Calcium Phosphate Precipitation in Aqueous Calcite Limestone Suspensions." *Journal of Environmental Quality* 9(4):641-644.

Brown, JL. 1981a. "Calcium Phosphate Precipitation: Effect of Common and Foreign Ions on Hydroxylapatite Crystal Growth." *Soil Science Society of America Journal* 45:482-486.

Brown, JL. 1981b. "Calcium Phosphate Precipitation: Identification of Kinetic Parameters in Aqueous Limestone Suspensions." *Soil Science Society of America Journal* 45:475-477.

Brown, WE, JP Smith, JR Lehr, and AW Frazier. 1962. "Crystalllographic and Chemical Relations Between Octacalcium Phosphate and Hydroxylapatite." *Nature* 196(4859):1050.

Brunauer, S, PH Emmett, and E Teller. 1938. "Adsorption of Gases in Multimolecular Layers." *Journal of the American Chemical Society* 60:300-319.

Chen, F, RC Ewing, and SB Clark. 1999. "The Gibbs Free Energies and Enthalpies of Formation of U^{6+} Phases: An Empirical Method of Prediction." *American Mineralogist* 84:650-664.

Christoffersen, J. 1980. "Kinetics of Dissolution of Calcium Hydroxylapatite, III. Nucleation-Controlled Dissolution of a Polydisperse Sample of Crystals." *Journal of Crystal Growth* 49:29-44.

Christoffersen, J, and MR Christoffersen. 1982. "Kinetics of Dissolution of Calcium Hydroxylapatite: V. The Acidity Constant for the Hydrogen Phosphate Surface Complex." *Journal of Crystal Growth* 57:21-26.

Chukhlantsev, VG, and SI Stepanov. 1956. "Solubility of Uranyl and Thorium Phosphates." *Russian Journal of Inorganic Chemistry* 1(3):478-484.

Conca, JL. 1996. *Phosphate-Induced Metal Stabilization*. StARS Program, National Center for Environmental Research, Office of Research and

Development, U.S. Environmental Protection Agency (EPA), Athens, Georgia.

de Leeuw, NH. 2004. "Resisting the Onset of Hydroxyapaite Dissolution Through the Incorporation of Fluoride." *Journal of Physical Chemistry B* 108:1809-1811.

DeFord, DH, RW Carpenter, and MW Finan. 1994. *300-FF-2 Operable Unit Technical Baseline Report*, BHI-00012, Rev. 00, Bechtel Hanford, Inc., Richland, WA.

Diaz, OA, KR Reddy, and PA Moore, Jr. 1994. "Solubility of Inorganic Phosphorous in Stream Water as Influenced by pH and Calcium Concentration." *Water Resources* 28(8):1755-1763.

DOE. 1995. *Remedial Investigation/Feasibility Study Report for the 300-FF-5 Operable Unit*, DOE/RL-94-85, Rev. 0, U.S. Department of Energy (DOE), Richland, WA.

DOE. 2005. *Limited Field Investigation Plan for the 300 Area Uranium Plume, 300-FF-5 Operable Unit, Hanford Site*, DOE/RL-2005-47, Rev. 1, published for the U.S. Department of Energy (DOE) by Pacific Northwest National Laboratory, Richland, WA.

DOE. 2005. *Work Plan for Phase III Feasibility Study, 300-FF-5 Operable Unit*, DOE/RL-2005-41, published for the U.S. Department of Energy (DOE) by Pacific Northwest National Laboratory, Richland, Washington.

Dorozhkin, SV. 1997a. "Acidic Dissolution Mechanism of Natural Fluorapatite, II. Nanolevel of Investigations." *Journal of Crystal Growth* 182:133-140.

Dorozhkin, SV. 1997b. "Surface Reactions of Apatite Dissolution." *Journal of Colloid and Interface Science* 191:489-497.

Dorozhkin, SV. 2002. "A Review on the Dissolution Models of Calcium Apatites." *Progress in Crystal Growth and Characterization of Materials* 44:45-61.

Eanes, ED, IH Gillessen, and AS Posner. 1965. "Intermediate States in the Precipitation of Hydroxylapatite." *Nature* 208:365-367.

Eanes, ED, and JL Meyer. 1977. "Maturation of Crystalline Calcium Phosphates in Aqueous Suspensions at Physiologic pH." *Calcified Tissue Research* 23(3):259.

Eanes, ED, and AS Posner. 1965. "Kinetics and Mechanism of Conversion of Noncrystalline Calcium Phosphate to Crystalline Hydroxylapatite." *Transactions New York Academy of Sciences* 28:233-241.

Eanes, ED, and AS Posner. 1970. "A Note on the Crystal Growth of Hydroxylapatite Precipitated from Aqueous Solutions." *Materials Research Bulletin* 5(6):377-384.

EPA. 1996. *Record of Decision for USDOE Hanford 300-FF-1 and 300-FF-5 Operable Units Remedial Actions*, Agreement Between the U.S. Department of Energy (DOE) and the U.S. Environmental Protection Agency (EPA), with concurrence of the Washington State Department of Ecology, EPA/ROD/R10-96/143, U.S. Environmental Protection Agency, Region 10, Seattle, WA.

Feenstra, TP, and PL de Bruyn. 1979. "Formation of Calcium Phosphates in Moderately Supersaturated Solutions." *Journal of Physical Chemistry* 83(4):475-479.

Ferguson, JF, D Jenkins, and J Eastman. 1973. "Calcium Phosphate Precipitation at Slightly Alkaline pH Values." *Water Pollution Control Federation* 45(4):620-631.

Ferguson, JF, D Jenkins, and W Stumm. 1970. "Calcium Phosphate Precipitation in Wastewater Treatment." *Chemical Engineering Progress Symposium Series* 107(67):279-287.

Finch, RJ. 1997. "Thermodynamic Stabilities of U(VI) Minerals: Estimated and Observed Relationships." *Material Research Society Symposium Proceedings* 465:1185-1192.

Finch, RJ, and T Murakami. 1999. "Systematics and Paragenesis of Uranium Minerals." In: PC Burns and RJ Finch, Eds., *Uranium: Mineralogy, Geochemistry and the Environment* (38:91-180). Washington, D.C.: Mineralogical Society of America.

Fuller, CC, JR Bargar, and JA Davis. 2003. "Molecular-Scale Characterization of Uranium Sorption by Bone Apatite Materials for a Permeable Reactive Barrier Demonstration." *Environmental Science and Technology* 37:4642-4649.

Fuller, CC, JR Bargar, JA Davis, and MJ Piana. 2002. "Mechanisms of Uranium Interactions with Hydroxylapatite: Implication for Groundwater Remediation." *Environmental Science and Technology* 36:158-165.

Garrels, RM, and CL Christ. 1965. *Solutions, Minerals and Equilibria.* New York: Harper and Row Publishing Company.

Garten, VA, and RB Head. 1966. "Homogeneous nucleation and the phenomenon of crystalloluminescence." *Philosophical Magazine* 14(132):1243.

Gauglitz, R, and M Holterdorf. 1992. "Immobilization of Heavy Metals by Hydroxylapatite." *Radiochimica Acta* 58(59):253-257.

Gerber, MS. 1992. *Past Practices Technical Characterization Study – 300 Area – Hanford Site*, WHC-MR-0388, Westinghouse Hanford Company, Richland, WA.

Giammar, DE. 2001. *Geochemistry of Uranium at Mineral-Water Interfaces: Rates of Sorption-Desorption and Dissolution-Precipitation Reactions.* Pasadena, CA: California Institute of Technology.

Gramin, P, JM Thomann, M Gumpper, and JC Voegel. 1989. "Dissolution kinetics of human enamel powder : I. Stirring effects and surface calcium accumulation." *Journal of Colloid Interfacial Science* 128:370.

Gramin P, JC Voegel, M Gumpper, and JM Thomann. 1987. "Surface properties and equilibrium kinetics of hydroxylapatite powder near the solubility equilibrium." *Journal of Colloid Interfacial Science* 118:147.

Grenthe, I, J Fuger, RJM Konings, RJ Lemire, AB Muller, C Nguyen-Trung, and H Wanner. 1992. *Chemical Thermodynamics of Uranium.* Amsterdam, Holland: Organisation for Economic Co-operation and Development (OECD), Nuclear Energy Agency.

Grossl, PR, and WP Inskeep. 1992. "Kinetics of Octacalcium Phosphate Crystal Growth in the Presence of Organic Acids." *Geochimica et Cosmochimica Acta* 56:1955-1961.

Guidry, MW, and FT Mackenzie. 2000. "Apatite Weathering and the Phanerozoic Phosphorus Cycle." *Geology* 28(7):631-634.

Guidry, MW, and FT Mackenzie. 2003. "Experimental Study of Igneous and Sedimentary Apatite Dissolution: Control of pH, Distance from Equilibrium, and Temperature on Dissolution Rates." *Geochimica et Cosmochimica Acta* 67(16):2949-2963.

Guillaumont, R, T Fanghanel, V Neck, J Fuger, DA Palmer, I Grenthe, and MH Rand. 2003. *Update on the Thermodynamics of Uranium, Neptunium, Plutonium, Americium and Technetium.* Amsterdam, Holland: Elsevier.

Harlov, DE, H-J Forster, and TG Nijland. 2002. "Fluid-Induced Nucleation of (Y+Ree)-Phosphate Minerals within Apatite: Nature and Experiment, Part I. Chlorapatite." *American Mineralogist* 87:245-261.

Hartman, MJ, LF Morasch, and WD Webber. 2007. *Hanford Site Groundwater Monitoring for Fiscal Year 2007*, PNNL-16346, Pacific Northwest National Laboratory, Richland, WA.

Icenhower, JP, and PM Dove. 2000. "The Dissolution Kinetics of Amorphous Silica into Sodium Chloride Solutions: Effects of Temperature and Ionic Strength." *Geochimica et Cosmochimica Acta* 64(24):4193-4203.

Icenhower, JP, PB McGrail, DM Strachan, RD Scheele, VL LeGore, EA Rodriguez, JL Steele, CF Brown, and MJ O'Hara. 2002. "Experimental Determination of the Dissolution Kinetics of Plutonium- and Uranium-Bearing Ceramics at 90°C." In: *Material Research Society Symposium*, Material Research Society Symposium Proceedings Vol 713 p. JJ11.38.

Icenhower, JP, PB McGrail, J Weber, Walter J., BD Begg, NJ Hess, EA Rodriguez, JL Steele, CF Brown, and MJ O'Hara. 2000. "Dissolution of $A_2ti_2o_7$ (a=Y^{3+}, Gd^{3+}, or Lu^{3+}) Pyrochlore by Experiment at pH=2, T=90°C: Evidence of Solubility Control Using a Linear Free Energy Model." In: *Material Research Society*, Material Research Society Symposium Proceedings Vol 713 p. JJ6.2.

Icenhower, JP, DM Strachan, MM Lindberg, EA Rodriguez, and JL Steele. 2003. *Dissolution Kinetics of Titanate-Based Ceramic Waste Forms: Results from Single-Pass Flow Tests on Radiation Damaged Specimens*, PNNL-14252, Pacific Northwest National Laboratory, Richland, WA.

Inskeep, WP, and JC Silvertooth. 1988. "Kinetics of Hydroxylapatite Precipitation at pH 7.4 to 8.4." *Geochimica et Cosmochimica Acta* 52:1883-1893.

Jenkins, D, JF Ferguson, and AB Menar. 1971. "Chemical Processes for Phosphate Removal." *Water Research* 5:369-389.

Jensen, MP, K Nash, JW Morse, EH Appelman, and MA Schmidt. 1996. "Immobilization of Actinides in Geomedia by Phosphate Precipitation." In: *ACS Symposium Series #651*, American Chemical Society, Washington D.C.

Kalmykov, SN, and GR Choppin. 2000. "Mixed $Ca^{2+}/UO_2^{2+}/CO_3^{2-}$ Complex Formation at Different Ionic Strengths." *Radiochimica Acta* 88:603-606.

Karpov, VI. 1961. "The Solubility of Triuranyl Phosphate." *Russian Journal of Inorganic Chemistry* 6:271-276.

Kato, K, Y Eika, and Y Ikada. 1997. "In Situ Hydroxylapatite Crystallization for the Formation of Hydroxylapatite/Polymer Composites." *Journal of Materials Science* 32:5533-5543.

Kohler, SJ, N Harouiya, C Chairat, and EH Oelkers. 2005. "Experimental Studies of Ree Fractionation During Water-Mineral Interaction: Ree Release Rates During Apatite Dissolution from pH 2.8 to 9.2." *Chemical Geology* 222:168-182.

Koutsoukos, PG, Z Amjad, MB Tomson, and GH Nancollas. 1980. "Crystallization of Calcium Phosphates: A Constant Composition Study." *Journal of American Chemical Society* 102(5):1553-1557.

Kura, G. 1987. "Alkaline Hydrolysis of Inorganic Cyclo-Polyphosphates." *Bulletin of Chemical Society of Japan* 60:2857-2860.

Kura, G, and T Tsukuda. 1993. "Effect of Copper(II), Nickel(II), and Aluminum(III) Ions on the Hydrolysis Rates of Inorganic Condensed Phosphate Oligomers." *Polyhedron* 12:865-870.

Langmuir, D. 1978. "Uranium Solution-Mineral Equilibria at Low Temperatures with Applications to Sedimentary Ore Deposits." *Geochimica et Cosmochimica Acta* 42:547-569.

Lee, SY, CW Francis, ME Timpson, and MP Elless. 1995. *Radionuclide Containment in Soil by Phosphate Treatment*, CONF-9503120-1, Oak Ridge National Laboratory, Oak Ridge, TN.

Lindsay, WL. 1979. "Phosphates." In: JWa Sons, Ed., *Chemical Equilibria in Soils* (163-209). Caldwell, NJ: The Blackburn Press.

Lindsay, WL, and EC Moreno. 1960. "Phosphate Phase Equilibria in Soils." *Soil Science Society of America Proceedings* 24:177-182.

Manecki, M, PA Maurice, and SJ Traina. 2000. "Uptake of Aqueous Pb by Cl^-, F^-, and OH^- Apatites: Mineralogic Evidence for Nucleation Mechanism." *American Mineralogist* 85:932-942.

McGrail BP, and SV Mattigod. 1999. *Laboratory Directed Research and Development Annual Report*, PNNL-12123, Pacific Northwest National Laboratory, Richland, WA.

McGrail, PB, WL Ebert, AJ Bakel, and DK Peeler. 1997. "Measurement of Kinetic Rate Law Parameters on a Na-Ca-Al Borosilicate Glass for Low-Activity Waste." *Journal of Nuclear Materials* 249:175-189.

McGrail, PB, and DK Peeler. 1995. *Evaluation of the Single-Pass Flow-Through Test to Support a Low-Activity Waste Specification*, PNL-10746, Pacific Northwest Laboratory, Richland, WA.

Misra, DN. 1991. "Adsorption of Low Molecular Weight Poly(Acrylic Acid) on Hydroxylapatite: Role of Molecular Association and Apatite Dissolution." *Langmuir* 7:2422-2424.

Moore, RC, C Sanchez, J Schelling, J Jones, DR Anderson, F Salas, D Lucero, and K Holt. 2001. *Bench-Scale Testing of In-Situ Formation of Apatite in Hanford Soils for Sorption of Uranium and Technetium*, SAND2001-3001, Sandia National Laboratories, Albuquerque, NM.

Moreno, EC, and K Varughese. 1981. "Crystal Growth of Calcium Apatites from Dilute Solutions." *Journal of Crystal Growth* 53:20-30.

Moskvin, AI, AM Shelyakina, and PS Perminov. 1967. "Solubility Product of Uranyl Phosphate and the Composition and Dissociation Constants of Uranyl Phosphato-Complexes." *Russian Journal of Inorganic Chemistry* 12(12):1756-1760.

Murakami, T, T Ohnuki, H Isobe, and T Sato. 1997. "Mobility of Uranium During Weathering." *American Mineralogist* 82:888-899.

Nagy, KL. 1995. "Dissolution and Precipitation Kinetics of Sheet Silicates." In: AF White and SL Brantley, Eds., *Chemical Weathering Rates of Silicate Minerals* (31:173-23). Washington, D.C.: Mineralogical Society of America.

Nancollas, GH, Z Amjad, and P Koutsoukos. 1979. "Calcium Phosphates – Speciation, Sorption, Solubility, and Kinetics." In: EA Jenne, Ed., *Chemical*

Modeling in Aqueous Systems (93:475-497). Washington D.C.: American Chemical Society.

Nancollas, GH, and MS Mohan. 1970. "The Growth of Hydroxylapatite Crystals." *Archives of Oral Biology* 15:731-745.

Nancollas, GH, and BB Tomazic. 1974. "Growth of Calcium Phosphate on Hydroxylapatite Crystals: Effect of Supersaturation and Ionic Medium." *Journal of Physics and Chemistry* 78:2218-2225.

Nash, K. 2000. *Organophosphorous Reagents in Actinide Separations: Unique Tools for Production, Cleanup and Disposal*, ANL/CHM/CP-100858, Argonne National Laboratory, Argonne, IL.

Nash, K, EJ Jensen, and MA Schmidt. 1998a. "In-Situ Actinide Immobilization of Actinides for Groundwater Cleanup: Laboratory Demonstration with Soil from the Fernald Environmental Management Project." In: WW Schultz and NJ Lombardo, Eds., *Science and Technology for Disposal of Radioactive Tank Wastes* (507-518). New York: Plenum Press.

Nash, K, MP Jensen, and MA Schmidt. 1997. "Actinide Immobilization in the Subsurface Environment by In-Situ Treatment with a Hydrolytically Unstable Organophosphorous Complexant: Uranyl Uptake by Calcium Phytate." In: *International Conference on Actinides*, Argonne National Lab, Argonne, IL ANL/CHM/CP--92802; CONF-970907, Argonne national laboratory, Argonne, IL.

Nash, K, MP Jensen, and MA Schmidt. 1998b. "Actinide Immobilization in the Subsurface Environment by In-Situ Treatment with a Hydrolytically Unstable Organophosphorous Complexant: Uranyl Uptake by Calcium Phytate." *Journal of Alloys and Compounds* 271-273:257-261.

Nash, K, LR Morse, MP Jensen, EH Appelman, MA Schmidt, S Friedrich, M Redko, and JJ Hines. 1999. *Water-Soluble Organophosphorous Reagents for Mineralization of Heavy Metals*, ANL/CHM/CP-98479, Argonne National Laboratory, Argonne, IL.

Nguyen, SN, RJ Silva, HC Weed, and JE Andrews. 1992. "Standard Gibbs Free Energies of Formation at the Temperature 303.15 K of Four Uranyl Silicates: Soddyite, Uranophane, Sodium Boltwoodite, and Sodium Weeksite." *Journal of Chemical Thermodynamics* 25:359-376.

O'Hare, PAG, J Boerio, and HR Hoekstra. 1976. "Thermochemistry of Uranium Compounds: Vii. Solution Calorimetry of Alpha and Beta-Na_2UO_4, Standard Enthalpy of Formation of Beta-Na_2UO_4 and the Enthalpy of the Alpha to Beta Transition at 298.15 K." *Journal of Chemical Thermodynamics* 8:845-855.

O'Hare, PAG, BM Lewis, and SN Nguyen. 1988. "Thermochemistry of Uranium Compounds Xvii. Standard Molar Enthalpy of Formation at 298.15 K of

Dehydrated Schoepite UO_3*0.9H_2O. Thermodynamics of (Schoepite+Dehydrated Scheopite+Water)." *Journal of Chemical Thermodynamics* 20:1287-1296.

Patton GP, BL Tiller, EJ Antonio, TM Poston, and SP Van Verst. 2003. *Survey of Radiological and Chemical Contaminants in the Near-Shore Environment at the Hanford Site 300 Area*, PNNL-13692, Rev. 1, Pacific Northwest National Laboratory, Richland, WA

Pekarek, V, and V Vesely. 1965. "A Study on Uranyl Phosphates, II. Sorption Properties of Some 1- to 4-Valent Cations on Uranyl Hydrogen Phosphate Heated to Various Temperatures." *Journal of Inorganic and Nuclear Chemistry* 27:1151-1158.

Peterson, RE (editor), EJ Freeman, CJ Murray, RE Peterson, PD Thorne, MJ Truex, VR Vermeul, MD Williams, SB Yabusaki, JM Zachara, JL Lindberg, and JP McDonald. 2005. *Contaminants of Potential Concern in the 300-FF-5 Operable Unit: Expanded Annual Groundwater Report for FY 2004*, PNNL-15127, Pacific Northwest National Laboratory, Richland, WA.

Pierce, EM, JP Icenhower, RJ Serne, and JG Catalano. 2005. "Experimental Determination of UO_2 (cr) Dissolution Kinetics: Effects of Solution Saturation State and pH." *Journal of Nuclear Materials* 345(2005):206-218.

Putnis, A, and CV Putnis. 2007. "The Mechanism of Reequilibration of Solids in the Presence of a Fluid Phase." *Journal of Solid State Chemistry* 180:1783-1786.

Renden-Angeles, JC, K Yanagisawa, N Ishizawa, and S Oishi. 2000. "Topotaxial Conversion of Chlorapatite and Hydrocyapatite to Fluorapatite by Hydrothermal Ion Exchange." *Chemistry of Materials* 12:2143-2150.

Schaad, P, F Poumier, JC Voegel, and P Gramin. 1997. "Analysis of Calcium Hydroxylapatite Dissolution in Non-Stoichiometric Solutions." *Colloids and Surfaces A: Physicochemical and Engineering Aspects* 121:217-228.

Schaad, P, J-M Thomann, J-C Voegel, and P Gramin. 1994. "Inhibition of Dissolution of Hydroxylapatite Powder by Adsorbed Anionic Polymers." *Colloids and Surfaces A: Physicochemical and Engineering Aspects* 83:285-292.

Scheyer, JM, and CF Baes. 1954. "The Solubility of Uranium (VI) Orthophosphates in Phosphoric Acid Solutions." *American Chemical Society Journal* 76:354-357.

Schmid, LA, and RR McKinney. 1969. "Phosphate Removal by a Lime-Biological Treatment Scheme." *Water Pollution Control Federation* 41(7):1258-1276.

Seaman, JC, JS Arey, and PM Bertsch. 2001. "Immobilization of Nickel and Other Metals in Contaminated Sediments by Hydroxylapatite Addition." *Journal of Environmental Quality* 30:460-469.

Sergeyeva, EI, AA Nikitin, IL Khodakovkiy, and GB Naumov. 1972. "Experimental Investigation of Equilibria in the System UO_3-CO_2H_2O in 25 - 200°C Temperature Interval." *Geochemistry International* 9:900-910.

Serne, RJ, WJ Martin, and VL LeGore. 1995. *Leach Test of Cladding Removal Waste Grout Using Hanford Groundwater*, PNL-10745, Pacific Northwest Laboratory, Richland, WA.

Shen, C, and D Dyroff. 1966. "Hydrolytic Degradation of Sodium Tripolyphosphate in Concentrated Solutions and in the Presence of Foreign Ions." *Industrial and Engineering Chemistry Product Research and Development* 5:97-100.

Shen, CY, and FW Morgan. 1973. "Hydrolysis of Phosphorous Compounds." In: EJ Griffith, A Beeton, JM Spencer, and DT Mitchell, Eds., *Environmental Phosphorous Handbook.* New York: John Wiley and Sons 241.

Sowder, AG, SB Clark, and RA Fjeld. 2000. "Dehydration of Synthetic Autunite Hydrates." *Radiochimica Acta* 88:533-538.

Stolyarova, TA. 1999. "Deficiency of Volatile Components in the Composition of Apatite." *Doklady Earth Sciences* 8(1):36.

Termine, JD, RA Peckauskas, and AS Posner. 1970. "Calcium Phosphate Formation in Vitro II. Effects of Environment on Amorphous-Crystalline Transformation." *Archives of biochemistry and biophysics* 140:318-325.

Termine, JD, and AS Posner. 1970. "Calcium Phosphate Formation in Vitro I. Factors Affecting Initial Phase Separation." *Archives of biochemistry and biophysics* 140:307-317.

Thomann, JM, JC Voegel, and P Gramin. 1990. "Kinetics of dissolution of calcium hydroxylapatite powder, III: pH and sample conditioning effects." *Calcified Tissue International* 46:121.

Tomazic, BB, and GH Nancollas. 1975. "The Seeded Growth of Calcium Phosphates. Surface Characterization and the Effect of Seed Material." *Journal of Colloid and Interface Science* 50:451-461.

Valocchi, AJ. 1984. "Describing the Transport of Ion-Exchanging Contaminants Using an Effective K_d Approach." *Water Resources Research* 20(4):499-503.

Valocchi, AJ. 1985. "Validity of the Local Equilibrium Assumption for Modeling Sorbing Solute Transport Through Homogeneous Soils." *Water Resources Research* 21(6):808-820.

Valsami-Jones, E, KV Ragnarsdottir, A Putnis, D Bosboach, AJ Kemp, and G Cressey. 1998. "The Dissolution of Apatite in the Presence of Aqueous Metal Cations at pH 2-7." *Chemical Geology* 151:215-233.

van Cappellen, P, and RA Berner. 1991. "Fluorapatite Crystal Growth from Modified Seawater Solutions." *Geochimica et Cosmochimica Acta* 55:1219-1234.

Vermuel, VR, JS Fruchter, DM Wellman, BA Williams, and MD Williams. 2006. *Site Characterization Plan: Uranium Stabilization through Polyphosphate Injection – 300 Area Uranium Plume Treatability Demonstration Project*, PNNL-16008, Pacific Northwest National Laboratory, Richland, WA.

Vesely, V, V Pekarek, and M Abbrent. 1965. "A Study on Uranyl Phosphates: III. Solubility Products of Uranyl Hydrogen Phosphate, Uranyl Orthophosphates and Some Alkali Uranyl Phosphates." *Journal of Inorganic and Nuclear Chemistry* 27:1159-1166.

Vochten, R. 1990. "Transformation of Cherikovite and Sodium Autunite into Lehnerite." *American Mineralogist* 75:221-225.

Waichler, SR, and SB Yabusaki. 2005. *Flow and Transport in the Hanford 300 Area Vadose Zone-Aquifer-River System*, PNNL-15125, Pacific Northwest National Laboratory, Richland, WA.

Watanabe, M, S Sato, and H Saito. 1975. "The Mechanism of the Hydrolysis of Condensed Phosphates, III. The Mechanism of the Hydrolysis of Trimeta- and Tetrametaphosphates." *Bulletin of the Chemical Society of Japan* 48:3593-3597.

Wazer, JRV. 1958. *Phosphorus and Its Compounds*. Interscience Publishers, Inc., New York.

Wazer, JRV, and E Griffith. 1955. "Structure and Properties of the Condensed Phosphates, VII. Hydrolytic Degradation of Pyro- and Tripolyphosphate." *Journal of American Chemical Society* 77:287-291.

Wellman, DM, JG Catalano, JP Icenhower, and AP Gamerdinger. 2005a. "Synthesis and Characterization of Sodium Meta-Autunite, $Na_2[(UO_2)(PO_4)]_2 \cdot 3H_2O$." *Radiochimica Acta* 93:1-7.

Wellman, DM, JI Icenhower, EM Pierce, BK McNamara, SD Burton, KN Geiszler, SR Baum, and BC Butler. 2005b. "Polyphosphate Amendments for In Situ Immobilization of Uranium Plumes." In: *Third International Conference on Remediation of Contaminated Sediments*, New Orleans, LA: Battelle Press.

Wellman, DM, JP Icenhower, and WJ Weber. 2005c. "Elemental Dissolution Study of Pu-Bearing Borosilicate." *Journal of Nuclear Materials* 340:149-162.

Wellman, DM, JS Fruchter, and VR Vermuel. 2006a. *Experimental Plan: Uranium Stabilization Through Polyphosphate Injection – 300 Area Uranium Plume Treatability Demonstration Project*, PNNL-16101, Pacific Northwest National Laboratory, Richland, WA.

Wellman, DM, JP Icenhower, AP Gamerdinger, and SW Forrester. 2006b. "Effects of pH, Temperature, and Aqueous Organic Material on the Dissolution Kinetics of Meta-Autunite Minerals, $(Na, Ca)_{2-1}[(UO_2)(PO_4)]_2 \cdot 3H_2O$." *American Mineralogist* 91:143-158.

Wellman, DM, JP Icenhower, and AT Owen. 2006c. "Comparative Analysis of Soluble Phosphate Amendments for the Remediation of Heavy Metal Contaminants: Effect on Sediment Hydraulic Conductivity." *Environmental Chemistry* 3:219-224.

Wellman, DM, EM Pierce, EL Richards, BC Butler, KE Parker, JN Glovack, SD Burton, SR Baum, ET Clayton, and EA Rodriguez. 2007a. *Interim Report: Uranium Stabilization Through Polyphosphate Injection – 300 Area Uranium Plume Treatability Demonstration Project*, PNNL-16683, Pacific Northwest National Laboratory, Richland, WA.

Wellman, DM, EM Pierce, and MM Valenta. 2007b. "Efficacy of Soluble Sodium Tripolyphosphate Amendments for the In Situ Immobilization of Uranium." *Environmental Chemistry* 4:293-300.

Wellman, DM, EL Richards, JN Glovack, and KE Parker. (*in press.*) "Sequestration of uranium (VI) in the presence of hydroxylapatite under dynamic geochemical conditions." *Environmental Chemistry*.

Wieker, W, and E Thilo. 1960. "The Catalytic Effect of Metal Ions on the Degradation of Polyphosphate in Aqueous Solutions." *Z. Anorg. Allgem. Chem.* 306:48-62.

Willard, J, T Farr, and J Hatfield. 1975. "Hydrolysis of Ammonium Pyro-, Tripoly-, and Tetrapolyphosphate at 25° and 50°C." *Journal of Chemical and Engineering Data* 20:276-283.

Williams, BA, CF Brown, W Um, MJ Nimmons, RE Peterson, BN Bjornstad, RJ Serne, FA Spane, and ML Rockhold. 2007. *Limited Field Investigation Report for Uranium Contamination in the 300 Area, 300-FF-5 Operable Unit, Hanford Site, Washington*, PNNL-16435, Pacific Northwest National Laboratory, Richland, WA.

Wolery, TJ. 1992. *Eq3nr, a Computer Program for Geochemical Aqueous Speciation-Solubility Calculations: Theoretical Manual, User's Guide, and Related Documentation (Version 7.0)*, UCRL-MA-110662, Lawrence Livermore National Laboratory, Livermore, CA.

Wolery, TJ, and SA Daveler. 1992. *Eq6, a Computer Program for Reaction Path Modeling of Aqueous Geochemical Systems: Theoretical Manual, User's Guide, and Related Documentation (Version 7.0)*, UCRL-MA-110662 PT IV, Lawrence Livermore National Laboratory, Livermore, CA.

Wolery, TW, and RL Jarek. 2003. *Eq3/6, Theoretical Manual, User's Guide, and Related Documentation (Version 8.0)*, Sandia National Laboratory, Albuquerque, NM.

Wright, J, LM Peurrung, TE Moody, JL Conca, X Chen, PP Didzerekis, and E Wyse. 1995. *In Situ Immobilization of Heavy Metals in Apatite Mineral Formulations*, Pacific Northwest Laboratory, Richland, WA.

Yanagisawa, K, JC Renden-Angeles, N Ishizawa, and S Oishi. 1999. "Topotaxial Replacement of Chlorapatite by Hydroxylapatite During Hydrothermal Ion Exchange." *American Mineralogist* 84:1861-1869.

INDEX

C

D

E

F

G

H

N

O

P

R

S

T

U

V

W

X

Y